FOUNDATION
GCSE Mathematics for Edexcel

HOMEWORK BOOK

JEAN LINSKY

SERIES EDITOR: ALAN SMITH

Hodder Murray
www.hoddereducation.co.uk

Acknowledgements

This high quality material is endorsed by Edexcel and has been through a rigorous quality assurance programme to ensure that it is a suitable companion to the specification for both learners and teachers. This does not mean that its contents will be used verbatim when setting examinations, nor is it to be read as being the official specification – a copy of which is available at www.edexcel.org.uk

Hodder Headline's policy is to use papers that are natural, renewable and recyclable products and made from wood grown in sustainable forests. The logging and manufacturing processes are expected to conform to the environmental regulations of the country of origin.

Orders: please contact Bookpoint Ltd, 130 Milton Park, Abingdon, Oxon OX14 4SB. Telephone: (44) 01235 827720. Fax: (44) 01235 400454. Lines are open 9 am to 5 pm, Monday to Saturday, with a 24-hour message answering service. Visit our website at www.hoddereducation.co.uk

Cover illustration © David Angel @ Début Art
Illustrations © Barking Dog
Typeset in Great Britain by Tech-Set Ltd
Printed and bound by Martins The Printers, Berwick-upon-Tweed.

A catalogue record for this title is available from the British Library.

ISBN-10: 0340 913 630
ISBN-13: 978 0340 913 635

Contents

Introduction

This Homework Book has been written to support the Hodder Murray Student's Book for the two-tier Edexcel GCSE Mathematics (Foundation Tier) specification for first teaching from September 2006. The book has been written with reference to the linear specification, though it can also be used effectively to cover the required content for the modular course.

The sequence of the material mirrors that of the Student's Book, so that, for example, Exercise 5.2 of the Homework Book provides further practice questions matching the content of Exercise 5.2 in the Student's Book.

An approximate indication of the level of difficulty of each exercise is given by the following set of icons:

no icon   grade E or F
● the start of D grade questions
■ the start of C grade questions

More practice helps to reinforce the ideas you have learned and makes it easier to remember at a later stage. If, however, you do forget, further help is at hand. As well as the Student's Book, with this book there is also a Personal Tutor CD-ROM. This contains worked examples on key topics to help you recall what you may have forgotten.

Look out for this symbol **pt** next to an exercise; it tells you that there is a relevant worked example on the CD-ROM.

All the questions in this book use an icon to show you whether calculators are permitted or not:

 You can use a calculator.

 Do not use a calculator.

You should follow this advice carefully in order to build up the necessary mixture of calculator and non-calculator skills before the examination.

All of the content of this book has been checked very carefully against the new GCSE specification, to ensure that all examination topics are suitably covered. In addition, practice exam-style papers, with full mark schemes, are available in Hodder Murray's accompanying Assessment Pack.

Chapter 1

Number review

EXERCISE 1.1

1 Write these numbers in words:
  a) 329   **b)** 1632   **c)** 4 836 007   **d)** 700 508

2 Write these numbers in figures:
  a) three thousand, five hundred and twenty
  b) seven hundred and four
  c) nine and a half thousand
  d) one million, eight hundred and two thousand
  e) five hundred thousand, four hundred and nine
  f) twenty-three thousand, seven hundred and eleven

3 Write these numbers in order of size – start with the smallest number:
  a) 507, 570, 499, 58, 432
  b) 830, 803, 847, 874, 809
  c) 4728, 4900, 4782, 4280, 4903
  d) 23 927, 72 990, 20 999, 9965, 21 973
  e) 1 759 400, 1 489 630, 1 081 927, 2 000 913, 1 050 559
  f) 652 000, 613 028, 607 956, 613 110, 608 280

EXERCISE 1.2

1 Work out:
  a) 53 + 61   **b)** 508 + 371   **c)** 79 + 12
  d) 735 + 86   **e)** 612 + 19 + 173   **f)** 908 + 404 + 37

2 Jean went shopping for a fruit salad.
She wrote a list of what she had bought.

Work out the total cost of the fruit she bought.

Fruit bought	
Oranges	£2.89
Apples	£1.70
Bananas	63p
Pears	87p
Grapes	£3.05

3 Find the sum of the numbers 6 to 11

4 Add 1209 to 653

5 Increase 258 by 674

6 Mrs Daurat has five children. Their ages are:

Peter	46
Lian	27
David	32
Steve	43
Ben	29

Work out the total age of her children.

EXERCISE 1.3

1 Work out:
   a) $87 - 45$
   b) $623 - 411$
   c) $56 - 39$
   d) $725 - 58$
   e) $812 - 375$
   f) $1467 - 889$

2 Decrease 719 by 304   3 Subtract 279 from 813

4 What number needs to be added to 38 to make 100?

5 Four students sat a Mathematics examination.
   The table shows their total marks out of 300:

Graham	213
Keith	95
Julie	187
Uzma	261

   a) What is the difference between Graham's and Keith's marks?
   b) How many more marks did Uzma get than Graham?
   c) What is the difference between Julie's mark and Uzma's mark?
   d) What is the difference between the highest mark and the lowest mark?
   e) How many marks did the four students get in total?

EXERCISE 1.4

1 Copy and complete this times table grid:

×	1	3	4	6	8	9	10
1	1			6			
2			8			18	
4		12					
5	5					45	
7			28				
8	8						
9		27					90

2 Work out:
   a) 5×8
   b) 9×3
   c) 4×7
   d) 6×2
   e) 5×7
   f) 8×6
   g) 10×9
   h) 7×6
   i) $5 \times 2 \times 7$
   j) $4 \times 5 \times 6$
   k) $3 \times 2 \times 9$
   l) $1 \times 1 \times 8$

3 Find the missing values:
   a) $7 \times ? = 42$
   b) $6 \times ? = 48$
   c) $? \times 10 = 70$
   d) $? \times 5 = 45$
   e) $3 \times ? = 27$
   f) $? \times 4 = 20$
   g) $9 \times ? = 18$
   h) $1 \times ? = 1$
   i) $3 \times 2 \times ? = 30$
   j) $? \times 4 \times 2 = 64$

EXERCISE 1.5

1 Multiply the following numbers by 10:
  a) 8   **b)** 57   **c)** 60   **d)** 914
  e) 2.36   **f)** 7.7   **g)** 0.8   **h)** 0.45

2 Multiply the following numbers by 100:
  a) 4   **b)** 23   **c)** 86   **d)** 308
  e) 0.5   **f)** 5.36   **g)** 1.7   **h)** 0.09

3 Multiply the following numbers by 1000:
  a) 5   **b)** 90   **c)** 436   **d)** 807
  e) 1.936   **f)** 0.064   **g)** 0.51   **h)** 23.7

4 Work out:
  a) 6×30 **b)** 8×20 **c)** 5×70 **d)** 4×900
  e) 3×600 **f)** 7×200 **g)** 30×50 **h)** 800×40
  i) 70×20 **j)** 2×800 **k)** 50×60 **l)** 700×3

EXERCISE 1.6

1 Divide the following numbers by 10:
  a) 70   **b)** 630   **c)** 4500   **d)** 30 000
  e) 57   **f)** 91   **g)** 6.2   **h)** 0.46

2 Divide the following numbers by 100:
  a) 7000   **b)** 300   **c)** 58 700   **d)** 2 670 000
  e) 19   **f)** 47   **g)** 0.58   **h)** 9.123

3 Divide the following numbers by 1000:
  a) 2000   **b)** 245 000   **c)** 63 000   **d)** 3 700 000
  e) 900   **f)** 60   **g)** 81   **h)** 5.4

4 Work out:
  a) $4000 \div 50$ **b)** $420 \div 70$ **c)** $3500 \div 700$
  d) $28\,000 \div 400$ **e)** $3500 \div 50$ **f)** $48\,000 \div 600$
  g) $600 \div 20$ **h)** $6300 \div 900$ **i)** $240\,000 \div 3000$
  j) $56\,000 \div 80$

EXERCISE 1.7 pt

1 Work out:
  a) 15×26 **b)** 37×54 **c)** 29×17
  d) 163×23 **e)** 191×32 **f)** 271×59
  g) 135×16 **h)** 284×43 **i)** 167×84

2 A large company posted 174 parcels. Each parcel needed an 83p stamp. Work out the total cost of the stamps.

3 There are 23 biscuits in each packet of Eatsumore biscuits.
  a) Work out how many biscuits there are in 54 packets.

  Each packet of biscuits costs £1.36
  b) Work out the total cost of the 54 packets.

EXERCISE 1.8

1 Work out:
  a) 315 ÷ 5   **b)** 432 ÷ 8   **c)** 282 ÷ 6
  d) 882 ÷ 14   **e)** 221 ÷ 17   **f)** 377 ÷ 13
  g) 522 ÷ 29   **h)** 966 ÷ 42   **i)** 952 ÷ 56

2 Work out the following – write down the remainder in each question:
  a) 68 ÷ 7   **b)** 45 ÷ 4   **c)** 614 ÷ 17
  d) 463 ÷ 25   **e)** 379 ÷ 11   **f)** 381 ÷ 12

3 A box can hold 18 books.
  Work out how many boxes will be needed to hold 666 books.

4 A librarian has £489 to spend on dictionaries.
  Each dictionary costs £23.
  How many dictionaries can the librarian buy?

EXERCISE 1.9

1 Ahmed went to a cafe.

  He ordered:

   3 sausages
   2 fried eggs
   baked beans
   mushrooms

  He paid with a £5 note.
  Work out how much change he got.

MENU	
Sausages	42p each
Fishcakes	87p each
Fried eggs	35p each
Onions	28p
Mushrooms	47p
Baked beans	50p

2 Mrs Fox buys rulers for her class.
  Each ruler costs 17p
  How many rulers can Mrs Fox buy if she has £5.00?

3 A florist sells roses at 36p each.
  Nina buys 144 roses for a function.
  Work out the total cost of the 144 roses.

4 Mary is 153 cm tall and weighs 104 kg
  John is 23 cm taller than Mary but weighs 17 kg less than her.
  Work out John's height and weight.

5 Boris plants bulbs along the border of his garden.
  He plants the bulbs 23 cm apart.
  If the length of the border is 675 cm, work out how many bulbs he planted.

6 Sunita buys 27 postcards. Each postcard costs 16p
  How much change will Sunita get if she pays with a £10 note?

EXERCISE 1.10

Do not use your calculator for Questions 1–5.

1 Work out:
a) $5 + 7$
b) $-2 + 8$
c) $-7 + 4$
d) $-3 + 9$
e) $1 + -3$
f) $-7 + -2$

2 Work out:
a) $9 - 5$
b) $-8 - 1$
c) $-7 - 3$
d) $7 - -2$
e) $6 - -3$
f) $-5 - -2$

3 Work out:
a) $6 + -3$
b) $-5 - -1$
c) $-8 + -3$
d) $-2 - 7$
e) $-9 - -9$
f) $-4 + -3$

4 Write these numbers in order of size. Write the smallest number first.
a) $-5, \ 7, \ 3, \ -4, \ -1, \ 4$
b) $8, \ -6, \ 7, \ 6, \ 0, \ -2$
c) $-3, \ 5, \ -9, \ -4, \ 7, \ 3$
d) $1, \ 0, \ -1, \ -5, \ 3, \ -2$

5 On 1st January at noon the temperature in six cities was recorded.
a) Write down:
(i) the highest temperature
(ii) the lowest temperature.
b) Work out the difference in temperature between Moscow and New York.
c) Work out the difference in temperature between London and New York.

The temperature in Oslo was 12 °C lower than in Barcelona.
d) Write down the temperature in Oslo.

City	Temperature (°C)
Moscow	−23
London	3
Cape Town	21
New York	−4
Barcelona	7

You can use your calculator for Questions 6 and 7.

6 Six people were in a competition. Each round they were awarded plus (+) points if they did well and minus (−) points if they did badly. The table gives information about their points.

Copy and complete the table.

Competitor	Round 1	Round 2	Total
Andy	+23	−20	
Brian	+4	−7	
Carol	−6		−10
Deborah	−3		+4
Ed		+5	−1
Frances		−9	+8

7 Gary has £5 and earns £26
He owes his mum £32 and his dad £19
He buys some food for £8 and gives £3 to charity.
His grandmother gives him £50 for his birthday.
Work out how much money Gary will have left if he pays all his debts.

EXERCISE 1.11

Work out the following. Be careful not to mix up all the different rules you have learnt.

1 5×-4
2 -3×-10
3 -7×3

4 $-45 \div -9$
5 $70 \div -7$
6 8×-2

7 $-36 \div -4$
8 -4×-4
9 $-18 \div 6$

10 6×-7
11 -8×-5
12 11×-6

13 $48 \div -6$
14 $35 \div -7$
15 5×-9

16 $-81 \div 9$
17 $-3 + -7$
18 $-6 - -8$

19 $-24 \div -6$
20 7×-8
21 $-63 \div 9$

22 $-6 - -3$
23 $-1 + -7$
24 $7 + -9$

25 -6×3
26 $8 \div -2$
27 $6 + -6$

EXERCISE 1.12

1 Round these numbers to the nearest 10: **a)** 23 **b)** 151 **c)** 867 **d)** 405

2 Round these numbers to the nearest 100: **a)** 382 **b)** 849 **c)** 550 **d)** 6351

3 Round these numbers to the nearest 1000: **a)** 676 **b)** 8099 **c)** 549 **d)** 3520

4 Round these numbers to the nearest integer: **a)** 6.7 **b)** 23.4 **c)** 9.2 **d)** 5.813

5 Round these numbers to 1 significant figure:
   a) 28 **b)** 73 **c)** 267 **d)** 155
   e) 3792 **f)** 408 **g)** 6193 **h)** 8705

6 Round these numbers to 2 significant figures:
   a) 813 **b)** 506 **c)** 127 **d)** 981
   e) 6352 **f)** 4438 **g)** 1823 **h)** 7907

EXERCISE 1.13

1 By rounding to 1 significant figure, work out an approximate answer to:
   a) 213×57
   b) $738 - 497$
   c) $897 \div 305$

   d) $\dfrac{63 \times 27}{85}$
   e) $\dfrac{815 \times 28}{782 - 213}$
   f) $\dfrac{91 \times 387}{903 - 285}$

   g) $\dfrac{7.6 \times 2.81}{3.19 + 2.73}$
   h) $\dfrac{7.2(3.06 + 1.15)}{1.93 \times 6.8}$
   i) $\dfrac{5.95(9.93 - 4.17)}{8.05 - 3.72}$

2 A large box contains 189 books each costing £3.28
   a) Estimate the total paid for the box of books.
   A school orders 5 of these boxes of books.
   b) Estimate the total cost of all 5 boxes of books.

Chapter 2

Ratio and proportion

EXERCISE 2.1

1 Simplify the following ratios:
   a) $4:8$ **b)** $6:9$ **c)** $18:12$ **d)** $14:21$
   e) $25:10$ **f)** $21:28$ **g)** $9:27$ **h)** $48:16$

2 Complete these equivalent ratios:
   a) $1:5 = 5:?$ **b)** $3:4 = 12:?$ **c)** $2:7 = 6:?$
   d) $9:10 = ?:50$ **e)** $5:8 = ?:24$ **f)** $5:6 = ?:30$

3 Match together the equivalent ratios:

$6:15$	$4:8$	$2:5$	$5:2$	$3:6$
$20:8$	$4:10$	$1:2$	$15:6$	

EXERCISE 2.2

1 Divide £150 in the following ratios:
   a) $2:3$ **b)** $7:8$ **c)** $5:12:13$ **d)** $3:5:7$

2 Divide £120 in the following ratios:
   a) $1:5$ **b)** $3:7$ **c)** $3:4:5$ **d)** $7:9:14$

3 A 60 cm line is cut into 3 pieces in the ratio $2:3:7$
   Work out the length of each piece.

4 Pete and Lian share some nuts.
   Pete takes 5 nuts for every 3 nuts that Lian takes.
   There are 48 nuts.
   How many **more** nuts does Pete take than Lian?

5 In a necklace, the ratio of silver beads to gold beads is $7:2$
   The necklace has a total of 36 beads.
   How many gold beads are in the necklace?

6 A box of chocolates has 2 milk chocolates for every 3 plain chocolates.
   There are 40 chocolates in the box.
   Work out how many plain chocolates are in the box.

7 To make fruit squash you mix 1 part of concentrate with 5 parts of water.
   How much water is needed to make 18 litres of fruit squash?

EXERCISE 2.3

1 Jason buys 5 concert tickets for £80
How much would seven concert tickets cost?

2 Bhavana earns £9.90 for $1\frac{1}{2}$ hours of work.
Work out how much she would earn for:
a) 30 minutes   **b)** 4 hours.

3 Here are the ingredients for making a vegetable soup for 18 people:

      6 carrots
      3 onions
      2400 ml stock
      150 g lentils
      12 g thyme

Work out how much of each ingredient is needed to make vegetable soup for
a) 9 people
b) 12 people
c) 36 people.

4 A train has 8 coaches and can seat 240 people.
Work out how many passengers a train with 10 coaches can seat.

5 In 5 months Jo earns £10 800
How much will Jo earn in a year in the same job?

6 Which of these is the best buy?

15 exercise books price £3.00	24 exercise books price £5.40

7 Sandra goes to South Africa. The exchange rate is £1 = 12 Rand.
a) How many Rand does she get for £200?

Sandra buys a wood carving for 180 Rand.
b) How much does the wood carving cost in pounds (£)?

8 Here is a recipe for orange jelly for 12 people.

      10 teaspoons gelatine
      400 g sugar
      600 ml water
      1.2 litres orange juice
      2 tablespoons lemon juice

How much of each ingredient is needed to make orange jelly for:
a) 24 people
b) 18 people?

EXERCISE 2.4

1 An architect draws a floor plan of a small school. She uses a scale of 1 : 50

   a) Work out the measurements, in centimetres, of the following rooms on her plan:
   - **(i)** classroom: 5 m by 8 m
   - **(ii)** head teacher's office: 3 m by 4 m
   - **(iii)** dining room: 10 m by 12 m

   > Remember that
   > 1 m = 100 cm

   b) Work out the measurements, in metres, of the following rooms:
   - **(i)** toilets: 6 cm by 4 cm on the plan
   - **(ii)** hall: 8.5 cm by 9.8 cm on the plan
   - **(iii)** kitchen: 7.1 cm by 4.7 cm on the plan.

   > Remember the
   > scale is 1 : 50

2 A model aeroplane is made on a scale of 1 cm representing 6 m

   a) Write this scale as a ratio.

   b) Work out the length of a wing on the real aeroplane if the length of a wing on the model aeroplane is 5 cm

   c) The length of the real aeroplane is 90 m
   Work out the length of the model aeroplane.

3 A map has a scale of 1 : 50 000

   a) What distance (in kilometres) is represented by a map distance of:
   - **(i)** 4 cm
   - **(ii)** 20 cm
   - **(iii)** 85 mm
   - **(iv)** 6.8 cm?

   > Remember
   > 1 km = 1000 m
   >   = 100 000 cm

   b) What distance (in centimetres) on the map is represented by a real distance of:
   - **(i)** 5 km
   - **(ii)** 55 km
   - **(iii)** 7.5 km?

Chapter 3

Decimals

EXERCISE 3.1

1 Round these numbers to 1 decimal place:
  a) 5.26
  b) 2.849
  c) 3.28
  d) 12.438
  e) 7.081
  f) 23.127
  g) 19.45
  h) 10.792

2 Round these numbers to 2 decimal places:
  a) 3.333
  b) 8.279
  c) 6.544
  d) 16.826
  e) 7.007
  f) 4.3571
  g) 8.996
  h) 11.9519

3 Round 16.708 352 to:
  a) 1 decimal place
  b) 2 decimal places
  c) 3 decimal places
  d) 4 decimal places.

EXERCISE 3.2

1 Round these numbers to 2 significant figures:
  a) 7.83
  b) 9.054
  c) 6.183
  d) 0.3867
  e) 0.007215
  f) 0.0009162
  g) 0.04774
  h) 0.0005555

2 Round these numbers to 3 significant figures:
  a) 2.9473
  b) 17.2497
  c) 267.76
  d) 0.08356
  e) 0.00048235
  f) 0.003892
  g) 0.05897
  h) 26.0489

3 Round 69.38547 to:
  a) (i) 1 significant figure
  (ii) 1 decimal place
  b) (i) 2 significant figures
  (ii) 2 decimal places
  a) (i) 3 significant figures
  (ii) 3 decimal places
  d) (i) 4 significant figures
  (ii) 4 decimal places.

EXERCISE 3.3

1 Work out:
  a) 0.4 + 0.5 + 0.3
  b) 0.6 + 0.1
  c) 2.2 + 3.1 + 0.8
  d) 5.23 + 8.19
  e) 8.03 + 2.91
  f) 7.54 + 6.13

2 Work out:
  a) 2.54 + 0.2
  b) 9.1 + 6.62
  c) 5.283 + 2.15
  d) 14.38 + 2.7
  e) 19.03 + 4.235
  f) 8.29 + 3.6 + 1.03

3 Work out:
  a) 5.3 − 2.1
  b) 0.85 − 0.24
  c) 9.37 − 4.17
  d) 12.7 − 3.9
  e) 24.61 − 20.24
  f) 37.73 − 14.61 − 3.01

4 Work out:
  a) 19.23 − 14.7
  b) 7.37 − 2.1
  c) 54.17 − 27.1
  d) 0.835 − 0.64
  e) 7.8 − 2.39
  f) 37.2 − 25.56

EXERCISE 3.4

1 Work out:
  a) 7×0.2 **b)** 0.5×9 **c)** 0.1×8 **d)** 4×0.6
  e) 0.5×0.3 **f)** 0.8×0.1 **g)** 0.05×0.6 **h)** 0.09×0.02
  i) 0.04×0.07 **j)** 1.2×0.3 **k)** $0.2 \times 0.3 \times 0.5$ **l)** $2.1 \times 0.1 \times 0.3$

2 Work out:
  a) 6.27×0.3 **b)** 5.09×0.2 **c)** 2.8×3.6
  d) 5.7×1.4 **e)** 7.14×3.8 **f)** 6.71×9.2

3 Use the fact that $215 \times 38 = 8170$ to work out:
  a) 2.15×3.8 **b)** 21.5×0.38 **c)** 215×3.8 **d)** 0.215×38

4 Use the fact that $518 \times 74 = 38\,332$ to work out:
  a) 51.8×7.4 **b)** 518×0.74 **c)** 5.18×7.4 **d)** 0.518×7.4

EXERCISE 3.5

1 Work out:
  a) $9 \div 0.3$ **b)** $8 \div 0.4$ **c)** $7 \div 0.1$ **d)** $24 \div 0.08$
  e) $35 \div 0.05$ **f)** $28 \div 0.07$ **g)** $42 \div 0.06$ **h)** $100 \div 0.04$

2 Work out:
  a) $3.6 \div 0.3$ **b)** $0.4 \div 0.2$ **c)** $0.15 \div 0.3$ **d)** $5.4 \div 0.09$
  e) $0.64 \div 0.08$ **f)** $2.5 \div 0.5$ **g)** $0.49 \div 0.07$ **h)** $6.3 \div 0.09$

3 Use the fact that $306 \div 17 = 18$ to work out:
  a) $30.6 \div 17$ **b)** $306 \div 0.17$ **c)** $3.06 \div 1.7$ **d)** $0.306 \div 0.17$

4 Use the fact that $28 \times 64 = 1792$ to work out:
  a) **(i)** $1792 \div 64$ **(ii)** $179.2 \div 28$ **(iii)** $17.92 \div 2.8$
  b) **(i)** $1.792 \div 0.28$ **(ii)** $17.92 \div 0.64$ **(iii)** $1792 \div 6.4$

EXERCISE 3.6

Do not use your calculator for Questions 1–3.

1 What numbers are the arrows **a**, **b**, **c** and **d** pointing to?

2 What numbers are the arrows **a**, **b**, **c** and **d** pointing to?

3 a) Write down the number marked with an arrow.

b) Write down the number marked with an arrow.

c) Find the number 620 on the number line. On a copy, mark it with an arrow (↑).

d) Find the number 4.9 on the number line. On a copy, mark it with an arrow (↑).

4 a) Write down the number marked with an arrow.

b) Write down the number marked with an arrow.

c) Find the number 57 on the number line. On a copy, mark it with an arrow (↑).

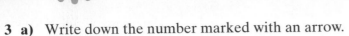

d) Find the number 3.9 on the number line. On a copy, mark it with an arrow (↑).

EXERCISE 3.7

Do not use your calculator for Questions 1–5.

1 Daniel buys 4 cans of cola at 82p each.
   a) How much does he spend?
   b) How much change does he get from £5?

2 A CD costs £7.50   Work out the cost of 9 of these CDs.

3 Gary is taking his family to the zoo.
Gary buys zoo tickets for 2 adults and 5 children.
a) How much does he spend on tickets?

Gary buys 7 sandwiches, 3 fruit juices and 5 chocolates.

Sandwiches	£2.20
Fruit juice	£1.34
Chocolates	£0.57

ZOO	
Adults	£9.30
Children	£6.00

b) How much does Gary spend on food and drinks?
c) How much does the zoo trip cost altogether?

4 Sunita orders soup, a burger and an ice cream.
Work out how much change she should get from £4

Menu	
Soup	86p
Burger	£1.64
Ice cream	60p

5 Will has twenty pounds and thirty-eight pence.
a) Write down the amount of money Will has in figures.
b) Will spends £8.53   How much money does he have left?

 You may use your calculator for Questions 6–8.

6

> *Haver Taxis Charges*
> £35.20 for the first 50 km
> £4.10 for each 10 km after this

Mia travels 100 km in a Haver taxi.
a) Work out the total amount she has to pay.

Fred travels in a Haver taxi. He pays £68
b) How far did Fred travel in the Haver taxi?

7 Suzy hires a car. She pays:
        £43.80 for the first day and
        £32.50 for each extra day.
Suzy hires a car for 1 week.
a) How much does she pay?

Kevin pays £433.80 to hire a car.
b) For how many days does he hire a car?

8 Jean went to the supermarket. She bought:
        2 cartons of soup at £1.12 per carton
        $1\frac{1}{2}$ kg of salmon at £14.10 per kg
        4 kg potatoes at £1.07 per kg
        3 melons at £2.16 each.

Work out how much change Jean should get from £40

Chapter 4

Fractions

EXERCISE 4.1

1 Write the following fractions as decimals:

a) $\frac{1}{4}$ b) $\frac{3}{8}$ c) $\frac{2}{5}$ d) $\frac{1}{3}$ e) $\frac{9}{100}$ f) $\frac{13}{20}$ g) $\frac{5}{6}$ h) $\frac{7}{12}$

2 Write the following numbers in order of size from smallest to biggest:

a) $\frac{2}{3}$, 0.61, $\frac{3}{5}$, 0.67, 0.605   b) 0.22, $\frac{1}{5}$, $\frac{1}{6}$, 0.18, $\frac{2}{7}$

c) $\frac{17}{20}$, 0.83, 0.819, $\frac{4}{5}$, 0.851   d) 0.38, $\frac{4}{11}$, 0.4, 0.3, $\frac{1}{3}$

EXERCISE 4.2

1 Match together equivalent fractions. | $\frac{3}{5}$ | $\frac{30}{48}$ | $\frac{21}{35}$ | $\frac{6}{27}$ | $\frac{10}{45}$ | $\frac{9}{15}$ | $\frac{20}{32}$ | $\frac{2}{9}$ | $\frac{5}{8}$ |

Do not use your calculator for Questions 2 and 3.

2 Write these fractions in their simplest form:

a) $\frac{24}{72}$ b) $\frac{42}{84}$ c) $\frac{8}{20}$ d) $\frac{5}{35}$ e) $\frac{9}{27}$ f) $\frac{50}{175}$ g) $\frac{72}{96}$ h) $\frac{39}{45}$

3 Find the missing values in these equivalent fractions:

a) $\frac{1}{5} = \frac{?}{10} = \frac{3}{?} = \frac{?}{25} = \frac{?}{55}$ b) $\frac{3}{4} = \frac{9}{?} = \frac{12}{?} = \frac{?}{40} = \frac{?}{60}$

c) $\frac{2}{3} = \frac{8}{?} = \frac{?}{24} = \frac{24}{?} = \frac{?}{60}$ d) $\frac{7}{10} = \frac{?}{20} = \frac{?}{30} = \frac{49}{?} = \frac{77}{?}$

4 Write these fractions in order of size:

a) $\frac{3}{4}$, $\frac{5}{8}$, $\frac{7}{16}$ b) $\frac{2}{5}$, $\frac{7}{10}$, $\frac{3}{10}$ c) $\frac{1}{4}$, $\frac{1}{6}$, $\frac{1}{3}$, $\frac{3}{24}$, $\frac{3}{8}$

d) $\frac{1}{12}$, $\frac{2}{3}$, $\frac{3}{5}$, $\frac{7}{10}$, $\frac{17}{30}$ e) $\frac{2}{3}$, $\frac{3}{4}$, $\frac{3}{6}$, $\frac{5}{12}$ f) $\frac{3}{10}$, $\frac{31}{100}$, $\frac{297}{1000}$

EXERCISE 4.3

1 Write the following decimals as fractions in their simplest form:

a) 0.1   b) 0.3   c) 0.25   d) 0.125   e) 0.86   f) 0.44   g) 0.325   h) 0.92

EXERCISE 4.4

1 Work out the following. Give your answer as a fraction in its simplest form:

a) $\frac{1}{5} + \frac{2}{5}$ b) $\frac{1}{8} + \frac{1}{8}$ c) $\frac{3}{11} + \frac{5}{11}$

d) $\frac{2}{7} + \frac{1}{7} + \frac{3}{7}$ e) $\frac{2}{15} + \frac{7}{15} + \frac{4}{15}$ f) $\frac{3}{9} + \frac{2}{9}$

2 Work out the following. Give your answer as a fraction in its simplest form:

a) $\dfrac{7}{11} - \dfrac{2}{11}$

b) $\dfrac{11}{12} - \dfrac{5}{12} - \dfrac{1}{12}$

c) $\dfrac{6}{7} - \dfrac{4}{7}$

d) $\dfrac{8}{15} - \dfrac{2}{15}$

e) $\dfrac{18}{25} - \dfrac{11}{25}$

f) $\dfrac{3}{5} - \dfrac{2}{5}$

3 Work out the following. Give your answer as a fraction in its simplest form:

a) $\dfrac{5}{8} + \dfrac{1}{4}$

b) $\dfrac{7}{12} - \dfrac{1}{3}$

c) $\dfrac{11}{14} - \dfrac{3}{7}$

d) $\dfrac{2}{3} + \dfrac{2}{9}$

e) $\dfrac{17}{20} - \dfrac{3}{5}$

f) $\dfrac{1}{4} + \dfrac{7}{16}$

g) $\dfrac{19}{25} - \dfrac{3}{5}$

h) $\dfrac{19}{30} - \dfrac{1}{6}$

i) $\dfrac{3}{8} + \dfrac{1}{2}$

4 Work out the following. Give your answer as a fraction in its simplest form:

a) $\dfrac{1}{5} + \dfrac{1}{2}$

b) $\dfrac{3}{8} - \dfrac{1}{3}$

c) $\dfrac{1}{4} + \dfrac{3}{7}$

d) $\dfrac{5}{6} - \dfrac{2}{3}$

e) $\dfrac{7}{9} - \dfrac{3}{4}$

f) $\dfrac{1}{6} + \dfrac{2}{5}$

g) $\dfrac{3}{5} + \dfrac{1}{4}$

h) $\dfrac{5}{7} - \dfrac{1}{3}$

i) $\dfrac{2}{5} + \dfrac{2}{3}$

j) $\dfrac{2}{3} + \dfrac{1}{2} - \dfrac{3}{4}$

k) $\dfrac{3}{5} - \dfrac{1}{3} + \dfrac{1}{4}$

l) $\dfrac{1}{3} + \dfrac{1}{4} - \dfrac{1}{2}$

5 Sarah spends $\frac{3}{7}$ of her monthly earnings on rent.
She spends $\frac{1}{7}$ of her monthly earnings on food.
a) What fraction of her monthly earnings does Sarah spend on rent and food?
b) What fraction of her monthly earnings does Sarah have left?

6 Jill spends $\frac{2}{5}$ of her pocket money on entertainment.
She spends $\frac{1}{3}$ of her pocket money on a CD.
Jill saves the rest of her pocket money.
a) What fraction of her pocket money does Jill spend?
b) What fraction of her pocket money does Jill save?

EXERCISE 4.5

1 Write the following as top-heavy fractions:

a) $1\frac{3}{5}$ b) $2\frac{1}{4}$ c) $1\frac{5}{7}$ d) $4\frac{2}{3}$ e) $5\frac{1}{6}$ f) $3\frac{1}{3}$ g) $6\frac{1}{2}$ h) $2\frac{4}{9}$

2 Write the following as mixed numbers:

a) $\frac{9}{5}$ b) $\frac{11}{3}$ c) $\frac{5}{2}$ d) $\frac{7}{4}$ e) $\frac{13}{5}$ f) $\frac{22}{7}$ g) $\frac{35}{6}$ h) $\frac{89}{10}$

3 Work out the following. Give your answer as a fraction in its simplest form:

a) $3\frac{5}{7} - 2\frac{3}{7}$

b) $1\frac{1}{4} + 2\frac{1}{4}$

c) $4\frac{2}{9} + 3\frac{5}{9}$

d) $3\frac{2}{3} - 2\frac{1}{3}$

e) $2\frac{5}{6} - 1\frac{1}{6}$

f) $1\frac{3}{5} + 2\frac{2}{5}$

4 Work out the following. Give your answer as a fraction in its simplest form:

a) $2\frac{1}{4} + 1\frac{1}{3}$

b) $1\frac{5}{6} - 1\frac{2}{3}$

c) $3\frac{1}{2} + 1\frac{2}{5}$

d) $2\frac{5}{7} - 1\frac{1}{2}$

e) $2\frac{3}{5} - 1\frac{1}{4}$

f) $1\frac{2}{3} + 2\frac{2}{5}$

g) $2\frac{5}{7} - 1\frac{3}{4}$

h) $3\frac{7}{8} - 2\frac{1}{4}$

i) $1\frac{1}{6} + 3\frac{1}{4}$

EXERCISE 4.6

1 Work out the following. Give your answer as a fraction in its simplest form:

a) $\frac{3}{4} \times \frac{2}{5}$ b) $\frac{2}{3} \times \frac{5}{7}$ c) $\frac{1}{7} \times \frac{1}{4}$ d) $\frac{5}{8} \times \frac{3}{7}$ e) $\frac{2}{9} \times \frac{4}{5}$ f) $\frac{3}{11} \times \frac{4}{9}$

2 Work out the following. Give your answer as a fraction in its simplest form:

a) $\frac{3}{7} \div \frac{9}{10}$ b) $\frac{3}{5} \div \frac{6}{7}$ c) $\frac{2}{9} \div \frac{4}{7}$ d) $\frac{2}{3} \div \frac{8}{11}$ e) $\frac{5}{6} \div \frac{10}{17}$ f) $\frac{8}{9} \div \frac{9}{8}$

3 Work out the following. Give your answer as a fraction in its simplest form:

a) $\frac{1}{5} \times \frac{10}{11}$ b) $\frac{3}{7} \times \frac{5}{6}$ c) $\frac{9}{10} \div \frac{18}{7}$ d) $\frac{1}{3} \div \frac{5}{6}$ e) $\frac{5}{7} \times \frac{14}{15}$ f) $\frac{1}{8} \div \frac{1}{7}$

g) $\frac{1}{4} \div \frac{3}{10}$ h) $\frac{6}{35} \times \frac{14}{9}$ i) $\frac{4}{5} \times \frac{5}{6}$ j) $\frac{4}{5} \div \frac{5}{6}$ k) $\frac{1}{5} \times \frac{5}{6} \div \frac{1}{3}$ l) $\frac{1}{5} \times \frac{1}{3} \div \frac{5}{6}$

EXERCISE 4.7

Do not use your calculator for Questions 1–3.

1 Work out:

a) $\frac{1}{4}$ of 28   b) $\frac{1}{2}$ of 46   c) $\frac{1}{3}$ of 12   d) $\frac{1}{10}$ of 90   e) $\frac{1}{7}$ of 21   f) $\frac{1}{5}$ of 555

2 Work out:

a) $\frac{2}{3}$ of 21   b) $\frac{3}{4}$ of 40   c) $\frac{5}{7}$ of 35

d) $\frac{3}{5}$ of 30   e) $\frac{4}{11}$ of 88   f) $\frac{7}{9}$ of 63

g) $\frac{7}{10}$ of 60   h) $\frac{5}{6}$ of 24   i) $\frac{3}{8}$ of 32

3 At Southvale School $\frac{13}{20}$ of the students are female.

There are 1200 students at the school.

a) How many students are female?

b) What fraction of the students is male?

c) How many students are male?

You can use your calculator for Questions 4–6.

4

CECIL'S CYCLES
$\frac{1}{5}$ off
Normal price £380

BILLY'S BICYCLES
$\frac{2}{7}$ off
Normal price £420

Which bicycle is the cheapest? Show all your working.

5 A book normally costs £23.80

a) How much does it cost in the sale?

Henry buys a book for £12

b) What is the normal price of the book?

> **BOOK SALE**
> $\frac{1}{4}$ off all normal prices

6 A pair of rollerblades normally costs £56

a) How much does it cost in the sale?

Jan buys a pair of rollerblades for £30

b) What is the normal price of the rollerblades?

> **Rollerblade Sale**
> $\frac{2}{5}$ off all normal prices

Chapter 5

Percentages

EXERCISE 5.1

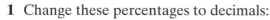

Do not use your calculator for Questions 1–4.

1 Change these percentages to decimals:
   a) 52%   **b)** 38%   **c)** 91%   **d)** 3%   **e)** 7%   **f)** 6.2%   **g)** 0.4%   **h)** 205%

2 Change these decimals to percentages:
   a) 0.37   **b)** 0.58   **c)** 0.6   **d)** 0.09   **e)** 0.04   **f)** 0.1   **g)** 3.85   **h)** 9.2

3 Change these fractions to percentages:
   a) $\frac{63}{100}$ **b)** $\frac{17}{100}$ **c)** $\frac{5}{100}$ **d)** $\frac{27}{50}$ **e)** $\frac{3}{10}$ **f)** $\frac{2}{5}$ **g)** $\frac{17}{25}$ **h)** $\frac{3}{4}$

4 Change these percentages to fractions. Write each answer in its simplest form.
   a) 65%   **b)** 90%   **c)** 36%   **d)** 4%   **e)** 12%   **f)** 340%   **g)** 12.5%   **h)** 50%

You can use your calculator for Questions 5 and 6.

5 Change these fractions to percentages:
   a) $\frac{28}{50}$ **b)** $\frac{15}{20}$ **c)** $\frac{57}{60}$ **d)** $\frac{12}{30}$ **e)** $\frac{24}{60}$ **f)** $\frac{22}{55}$ **g)** $\frac{12}{40}$ **h)** $\frac{36}{80}$

6 Copy and complete the following table.

Percentage	Decimal	Fraction
	0.25	
		$\frac{1}{2}$
10%		
0.3333	$\frac{1}{3}$	
0.9		
$66\frac{2}{3}\%$		
		$\frac{3}{4}$
80%		
	0.05	
20%		

EXERCISE 5.2

Do not use your calculator for Questions 1–3.

1 A school has 300 students in each of Years 9, 10 and 11.
The number of girls in each year group are:

    120 girls in Year 9
    210 girls in Year 10
    150 girls in Year 11

Write each of these as a percentage of the number of students in that year.

2 Maria buys a house for £240 000 and sells it three years later for £300 000
    a) How much profit has she made?
    b) What is her percentage profit?

3 Harry buys a book for £40 and sells it for £32
    a) How much money has he lost?
    b) What is his percentage loss?

You can use your calculator for Questions 4 and 5.

4 Jasmine gets these marks in her end of
  term tests.

    a) Write each of Jasmine's marks
        as a percentage.
    b) In which subject did she get
        her worst mark?
    c) In which subject did she get
        her best mark?

Subject	Mark
Business Studies	$\frac{22}{40}$
French	$\frac{36}{60}$
Maths	$\frac{93}{150}$
Religious Studies	$\frac{12}{30}$
Science	$\frac{132}{200}$
Spanish	$\frac{26}{50}$

5 Pete bought some items on the internet and then sold them at a garage sale.
  What percentage profit or loss has he made on each of these items?
    a) Necklace bought for £240, sold for £360
    b) Boots bought for £60, sold for £45
    c) Painting bought for £500, sold for £380
    d) A CD collection bought for £80, sold for £90

EXERCISE 5.3

1 Work out:
    a) 30% of 360   **b)** 20% of 45   **c)** 80% of 65   **d)** 7% of 300
    e) 25% of 420   **f)** 42% of 60   **g)** 10% of 810   **h)** 2% of 790
    i) 2.5% of 400   **j)** 17.5% of 1000   **k)** 150% of 6342   **l)** 130% of 500

2 Jon buys a video recorder.
How much VAT does he pay?

> **SPECIAL OFFER!**
> *Video Recorder*
> only £290 plus 17.5% VAT

3 In a Mathematics test 84% of the students passed.
  a) What percentage of the students did not pass?

200 students sat the test.
  b) How many students passed the test?

4

> **STEVE'S SAUSAGES**
> 72% beef

> **BERTIE'S BANGERS**
> 44 g out of every 60 g beef

Rob wants to buy sausages that contain the most beef.
Which sausages should he buy? Explain why.

5 Neil gets 2% commission for every house he sells.
Work out how much commission he gets when he sells houses for:
  a) £320 000   **b)** £65 500   **c)** £1 270 000

EXERCISE 5.4

1 Find   **(i)** 10%   **(ii)** 5%   **(iii)** 15% of:
  a) 300   **b)** 40   **c)** 1500   **d)** 76

2 Find 25% of:
  a) 300   **b)** 20   **c)** 6400   **d)** 48

3 Find:
  a) 50% of 146   **b)** 25% of 280   **c)** 5% of 60
  d) 7.5% of 200   **e)** 10% of 6530   **f)** 40% of 55
  g) 20% of 350   **h)** 65% of 2400   **i)** 70% of 700

4 Work out the VAT (at 17.5%) that needs to be paid on the following items.
  a)

> **TV set**
> *price: £250*

  b)

> **Oil painting**
> *price: £6200*

  c)

> **Electronic organiser**
> *price: £180*

5 A shop offers a discount of 15% to all of its customers.
How much is the discount on the following items?
  a)

> **Gold earrings**
> *price: £320*

  b)

> **Silver clock**
> *price: £680*

  c)

> **Gold pen**
> *price: £84*

EXERCISE 5.5

Do not use your calculator for Questions 1–3.

1. A choir had 50 members in January.
   By the next December its membership had increased by 24%
   How many members did the choir have in December?

2. Sheila went on a 3-month cruise.
   She weighed 74 kg at the start of the cruise.
   By the end of the cruise her weight had increased by 12%
   Work out how much Sheila weighed at the end of the cruise.

3. Scott invested £2300 at 3% simple interest for one year.
   How much does he have after one year?

You can use your calculator for Questions 4–6.

4. In a sale all items are reduced by 14%.
   What is the sale price for each of the following items?

Item	Price before the sale
Widescreen televisions	£800
DVD players	£64
Computers	£1200
Laptops	£950

5. Simon's restaurant bill is £64 plus a service charge of 12.5%
   How much must Simon pay?

6. Matt invests £7800 at 2.5% simple interest for one year.
   How much does he have after one year?

EXERCISE 5.6

1. Lily invests £6000 at 4% compound interest.
   How much does she have after: **a)** 1 year   **b)** 2 years   **c)** 3 years?

2. A £23 000 car depreciates by 15% every year.
   So its value **decreases** by 15% every year.
   How much is the car worth after: **a)** 1 year   **b)** 2 years   **c)** 3 years?

3. Myra invests £8000 at 6% compound interest.
   How much does she have after 3 years?

4. The value of a painting in an art gallery has increased by 20% each year for the last 4 years.
   It was worth £53 000, 4 years ago.
   How much is it worth today?
   Give your answer to the nearest thousand pounds.

Chapter 6

Powers and roots

EXERCISE 6.1

1 a) Write down the first six multiples of:
  (i) 9 **(ii)** 12 **(iii)** 18 **(iv)** 24 **(v)** 36
  b) Find the **least common multiple** of:
  (i) 12 and 18 **(ii)** 9 and 24 **(iii)** 18 and 24 **(iv)** 24 and 36

2 a) Find the factors of:
  (i) 15 **(ii)** 18 **(iii)** 24 **(iv)** 120
  b) Find the **highest common factor** of:
  (i) 15 and 18 **(ii)** 18 and 24 **(iii)** 18 and 120

3 Which of the following numbers are **not** prime? Give a reason for each answer.

  15 17 26 31 35 49 53 60 67

4 a) Which of the following numbers are prime?
  If it is not prime, give a number that goes into it exactly.
  7, 17, 27, 37, 47, 57, 67, 77, 87, 97
  b) Which of the following numbers are prime?
  If it is not prime, give a number that goes into it exactly.
  9, 19, 29, 39, 49, 59, 69, 79, 89, 99

EXERCISE 6.2

1 Square the following numbers:
  a) 4 **b)** 7 **c)** 8 **d)** 10 **e)** 11

2 Work out:
  a) 3^2 **b)** 5^2 **c)** 8^2 **d)** 6^2 **e)** 12^2

3 Write down the first 10 square numbers.

4 Find the square root of: **a)** 36 **b)** 81 **c)** 64 **d)** 100

5 Work out: **a)** $\sqrt{16}$ **b)** $\sqrt{25}$ **c)** $\sqrt{49}$ **d)** $\sqrt{1}$

6 2500 counters are arranged in a square.
  How many counters are along one side of the square?

7 a) (i) Square root 16 and then square your answer.
  (ii) Square root 36 and then square your answer.
  (iii) What do you notice?
  b) Work out: **(i)** $(\sqrt{7})^2$ **(ii)** $(\sqrt{11})^2$ **(iii)** $(\sqrt{20})^2$

8 Write down the whole number that is closest in value to:
  a) $\sqrt{24}$ **b)** $\sqrt{6}$ **c)** $\sqrt{97}$ **d)** $\sqrt{11}$

9 Estimate: **a)** $\sqrt{50}$ **b)** $\sqrt{108}$ **c)** $\sqrt{28}$ **d)** $\sqrt{5}$

EXERCISE 6.3

1 Work out:
  a) $\sqrt{361}$ **b)** $\sqrt{39.69}$ **c)** $\sqrt{0.09}$ **d)** $\sqrt{70.56}$

2 Work out the following – give your answers to 1 decimal place:
  a) 3.6^2 **b)** 9.7^2 **c)** 24.1^2 **d)** 43.3^2

3 Find the square root of:
  a) 1369 **b)** 1.44 **c)** 204.49 **d)** 21.16

4 Work out the following – give your answers to 3 significant figures:
  a) $3.8^2 + \sqrt{250}$ **b)** $\sqrt{52} - \sqrt{37}$ **c)** $21.9^2 - \sqrt{126}$

5 Work out the following – give your answers to 3 significant figures:
  a) $\dfrac{\sqrt{983}}{6.7^2}$ **b)** $\dfrac{4.1^2}{\sqrt{9.2}}$ **c)** $\dfrac{\sqrt{543}}{2.1^2}$

EXERCISE 6.4

1 Which of the following numbers are cube numbers?
  4   8   16   27   64

2 Work out:
  a) 1^3 **b)** 6^3 **c)** 10^3 **d)** 0.2^3

3 Work out:
  a) $\sqrt[3]{2744}$ **b)** $\sqrt[3]{125}$ **c)** $\sqrt[3]{0.001}$ **d)** $\sqrt[3]{4.913}$

EXERCISE 6.5

1 Write the following using indices:
  a) $2 \times 2 \times 2 \times 2 \times 2 \times 2$ **b)** $5 \times 5 \times 5 \times 5$
  c) $8 \times 8 \times 8 \times 8 \times 8 \times 8 \times 8 \times 8$ **d)** $1 \times 1 \times 1$
  e) $4.7 \times 4.7 \times 4.7 \times 4.7 \times 4.7$ **f)** $9.3 \times 9.3 \times 9.3 \times 9.3 \times 9.3 \times 9.3 \times 9.3$

2 Use your calculator to work out:
  a) 6^4 **b)** 8^3 **c)** 1^{23} **d)** 3^7 **e)** 2^9 **f)** 9^5

3 Use your calculator to work out:
  a) $\sqrt[5]{243}$ **b)** $\sqrt[9]{1}$ **c)** $\sqrt[8]{6561}$ **d)** $\sqrt[5]{32\,768}$ **e)** $\sqrt[4]{0.4096}$ **f)** $\sqrt[7]{128}$

4 a) Match together cards with the same answer.

$3^5 \times 3^3$	$3^6 \div 3^5$	3^3	3^8	$3^7 \times 3^2$	$3^7 \div 3^3$
3^1	3^5	3×3^4	$3^5 \div 3^2$	3^9	3^4

b) What do you notice?

EXERCISE 6.6

Do not use your calculator for Questions 1 and 2.

1 Write each of the following as a single power:
a) $7^5 \times 7^9$ b) $2^{10} \times 2^6$ c) $10^8 \div 10^2$ d) $5^{18} \div 5^6$
e) $8^4 \times 8^5$ f) $9^9 \times 9$ g) $15^{80} \div 15^{50}$ h) $37^{14} \div 37^7$
i) $2.6^3 \times 2.6^4$ j) $58^8 \times 58^8$ k) $\dfrac{6^{15}}{6^3}$ l) $\dfrac{9^{12}}{9^9}$

2 Write each of the following as a single power:
a) $\dfrac{7^3 \times 7^8}{7^6}$ b) $\dfrac{3^9 \times 3^5}{3^7}$ c) $\dfrac{8^2 \times 8^9}{8^4}$

You can use your calculator for Questions 3 and 4.

3 a) Write the following as a single power:
(i) $\dfrac{5^7}{5^6}$ (ii) $\dfrac{8^5}{8^4}$ (iii) $\dfrac{6^9}{6^8}$

b) Work out the value of:
(i) $\dfrac{5^7}{5^6}$ (ii) $\dfrac{8^5}{8^4}$ (iii) $\dfrac{6^9}{6^8}$

c) How else can you write x^1?

4 a) Write the following as a single power:
(i) $\dfrac{4^7}{4^7}$ (ii) $\dfrac{9^8}{9^8}$ (iii) $\dfrac{2^{10}}{2^{10}}$

b) Work out the value of:
(i) $\dfrac{4^7}{4^7}$ (ii) $\dfrac{9^8}{9^8}$ (iii) $\dfrac{2^{10}}{2^{10}}$

c) What does x^0 always equal?

EXERCISE 6.7

1 Write each of the following numbers as a product of prime factors.
a) 24   b) 63   c) 45   d) 120   e) 36
f) 84   g) 30   h) 420   i) 90

2 Find the **(i)** highest common factor and **(ii)** least common multiple of:
a) 36 and 45   b) 63 and 84   c) 90 and 420
d) 36 and 63   e) 120 and 420   f) 30 and 45
g) 84 and 420   h) 24 and 90   i) 24 and 36

EXERCISE 6.8

1 Write down the reciprocal of:
a) 3   b) 5   c) 9   d) 0.5
e) 0.125   f) $\dfrac{1}{7}$ g) $\dfrac{5}{8}$ h) $\dfrac{2}{3}$

2 a) Write down the reciprocal of:
(i) 5   (ii) 0.4   (iii) $\dfrac{7}{11}$

b) Write down the reciprocal of each of your answers to **a)**. What do you notice?

EXERCISE 6.9

1 Work out:

a) $6 + 3 \times 2$

b) $10 \div 2 + 3$

c) $(7 + 2) \times 2 + 5$

d) $3 \times 8 - 6 \times 2$

e) $\dfrac{9 \times 2}{5 + 1}$

f) $\dfrac{5 + 3^2}{2^3 - 1}$

g) $\dfrac{10 - 4 \times 2}{12 \div 6}$

h) $\dfrac{15 - 2 \times 3}{10 - (5 + 4)}$

2 Use your calculator to work out the following.
Give your answers to 1 decimal place:

a) $23 - \sqrt{37}$

b) $7 \times \sqrt{(26 \div 4)}$

c) $\dfrac{15.8 + 3.72}{8.9 - 4.01}$

d) $7.8^2 + 12.3$

e) $93.4 \div (2 + \sqrt{45})$

f) $\dfrac{17^2 - 13^2}{\sqrt{(8.2 - 3.56)}}$

g) $\dfrac{62.1 \times 23.4}{7.8^2 + \sqrt{37}}$

h) $\dfrac{\sqrt{2378}}{9.7^2}$

3 Insert brackets into the following sums to make them correct.
a) $6 + 9 \div 3 = 5$
b) $8 \times 6 + 3 \div 6 - 3 = 9$
c) $2.4 + 7.2 \times 3.7 - 1.6 = 20.16$

EXERCISE 6.10

1 Work out:
a) 10^3

b) 10^5

c) 10^9

2 Write the following as a power of 10:
a) $10\,000$

b) $100\,000\,000$

c) $1\,000\,000\,000\,000$

3 10^{100} is called a 'googol'.
One hundred googols is written as 1 followed by how many zeros?

4 Write each of the following: **(i)** using powers of 10; **(ii)** as an ordinary number.
a) 5 hundred thousand
b) 3 million
c) 70 thousand

5 Use your calculator to work out the following, giving your answer in standard index form:
a) $(2.7 \times 10^5) \times (3 \times 10^7)$
b) $(7.8 \times 10^{15}) \div (2 \times 10^{11})$
c) $(3.6 \times 10^8) \times (1.9 \times 10^{13})$
d) $(5.1 \times 10^{23}) \div (3 \times 10^{12})$

Chapter 7

Working with algebra

EXERCISE 7.1

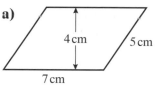

Do not use your calculator for Questions 1 and 2.

1 Use the formulae:

area = base × perpendicular height
perimeter = 2 × base + 2 × length of sloping side

to work out the area and perimeter of these parallelograms:

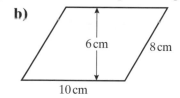

a) 4 cm, 5 cm, 7 cm

b) 6 cm, 8 cm, 10 cm

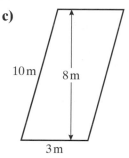

c) 10 m, 8 m, 3 m

2 Use the formula $T = H \times R$, where T is the total pay, H is the number of hours worked, and R is the hourly rate, to work out the total pay for the following people:
   a) June works 20 hours and is paid £9 per hour
   b) Marvin is paid an hourly rate of £5 and he works for 70 hours
   c) Fahir's hourly rate of pay is £6 and he works for 12 hours
   d) Beth only works for 90 minutes but her hourly rate is £22 per hour.

You can use your calculator for Questions 3–5.

3 A caterer uses the following formula to work out their customers' catering bills.

Total bill (in £)	=	Standard charge	+	number of people at the dinner × £23	+	number of helpers needed × £50

Jasmine had 16 people for dinner and needed 3 helpers. The standard charge was £40
   a) How much does Jasmine have to pay the caterer?

Mike had a large birthday party. There were 100 people at the party and 8 helpers were needed. The standard charge was £40
   b) Work out Mike's total bill.

4 The speed of a car is given by the formula:
   $$S = D \div T$$
   Find the exact value of S when:
   a) $D = 540$ and $T = 18$ **b)** $D = 2142$ and $T = 34$
   c) $D = 1288$ and $T = 23$ **d)** $D = 302.4$ and $T = 3.6$

5 The area, A, of a trapezium is found using the formula:

$$A = \frac{p + q}{2} \times h$$

where p and q are the lengths of the parallel sides and h is the distance between these sides.

Find A when:
a) $p = 5, q = 7$ and $h = 6$ **b)** $p = 6, q = 8$ and $h = 2$
c) $p = 12, q = 18$ and $h = 5$ **d)** $p = 10, q = 5$ and $h = 3.6$

EXERCISE 7.2

1 Write down expressions for the following:
  a) 3 more than x **b)** 8 more than w
  c) 5 less than v **d)** 23 multiplied by c
  e) r lots of 4   **f)** r squared
  g) $5a$ more than $3b$ **h)** 23 divided by d
  i) the sum of g and $2h$ **j)** p divided by q
  k) $3e$ all squared   **l)** t divided by u
  m) 5 lots of f squared   **n)** c squared multiplied by n cubed

2 Kevin has k sweets.
  Julie has 5 more sweets than Kevin.
  a) Write down an expression, in terms of k, for the number of sweets that Julie has.
  Peter has 8 more sweets than Julie.
  b) Write down an expression, in terms of k, for the number of sweets that Peter has.

  Joe has 3 times as many sweets as Peter.
  c) Write down an expression, in terms of k, for the number of sweets that Joe has.
  d) Kevin has 17 sweets. How many sweets does Joe have?

3 The base of a parallelogram is 7 cm more than its height, h.
  a) Write down an equation, in terms of h, for the base of the parallelogram.
  b) Write down an equation, in terms of h, for the area of the parallelogram.
  c) Work out the area of parallelogram when:
    (i) $h = 5$ cm   **(ii)** $h = 1$ cm   **(iii)** $h = 4$ cm

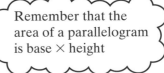

Remember that the area of a parallelogram is base $\times$ height

4 Write down an expression for the area of this square. x cm
  a) Nine of these squares are put together to make a large square:
  b) Write down an expression for the area of the large square.
  c) What is the length of one side of the large square?
  d) Explain how this shows that $9x^2 = (3x)^2$

EXERCISE 7.3

Do not use your calculator for Questions 1–3.

1 When $p = 2, q = 6, r = 0$ and $t = 3$ find the value of:
- **a)** $t + 6$
- **b)** $8 - p$
- **c)** $5q$
- **d)** $4p - t$
- **e)** $8r + q + p$
- **f)** $7p + t$
- **g)** q^2
- **h)** pq
- **i)** $2.5p + 6t$
- **j)** ptr
- **k)** q^p
- **l)** t^3

2 When $x = 3, y = 8$ and $z = 4$ find the value of:
- **a)** $5(x + z)$
- **b)** $(2z)^2$
- **c)** $2z^2$
- **d)** $x(y - z)$
- **e)** $10y^2$
- **f)** $z(2y - 2x)$
- **g)** $\dfrac{2y}{z}$
- **h)** $(x + y)^2$
- **i)** $\dfrac{x + y + z}{5}$

3 When $a - 9, b - 5$ and $c = -2$ find the value of:
- **a)** $b + c$
- **b)** $a + b^2$
- **c)** $c^2 + b$
- **d)** $\dfrac{a + b}{c}$
- **e)** $\dfrac{b - c}{a - 2}$
- **f)** $\dfrac{2a}{b + c}$

You can use your calculator for Questions 4–8.

4 When $c = 5, h = 7$ and $n = 3$, find the value of:
- **a)** $2h - 3n$
- **b)** $c^2 - 5n$
- **c)** $ch - 2hn$
- **d)** $(3c - 2h)^2$

5 When $e - 2, f - 4$ and $g = -3$, find the value of:
- **a)** $(e + f)^2$
- **b)** $3g + 2ef$
- **c)** $2f - g$
- **d)** $(3e - g)^2$

6 When $x = -6, y = -1$ and $z = 9$, find the value of:
- **a)** $y^2 + 2z$
- **b)** $xy + 2$
- **c)** $5(y + x)$
- **d)** $z(x - 10y)$
- **e)** $5y^8$
- **f)** $2x^2$
- **g)** $(z + x)^3$
- **h)** $7(2x + 3z)$
- **i)** $\dfrac{x + 2z}{2y}$
- **j)** $\dfrac{x^2}{yz}$

7 The velocity, v, of a particle is found using the formula:
$$v = u + at$$
where u is the initial velocity, t is the time and a is the acceleration.
- **a)** Use your calculator to work out the value of v when $a = 9.8, t = 3.57$ and $u = 2.43$. Write down all the figures on your calculator display.
- **b)** Round your answer to part **a)** correct to 3 significant figures.

8 The distance, s, travelled by a particle is given by the formula:
$$s = ut + \tfrac{1}{2}at^2$$
where u is the initial velocity, t is the time and a is the acceleration.
- **a)** Use your calculator to work out the value of s when $a = 8.5, t = 6.5$ and $u = 1.7$ Write down all the figures on your calculator display.
- **b)** Round your answer to part **a)** correct to an appropriate degree of accuracy.

EXERCISE 7.4

1 Simplify
  a) $a + a + a$
  b) $2b + 5b + 6b$
  c) $3c + c + 2c + c$
  d) $4d + 2d + 5d - 3d$
  e) $8e + e - 3e$
  f) $12f - f - 2f - 4f$
  g) $2g + 5g - 10g$
  h) $h - 5h$
  i) $2i - 4i - 6i + 5i$

2 Simplify:
  a) $a + 5 + a + 3$
  b) $2b + 7 + b + 3$
  c) $6c + 5 - 2c - 1$
  d) $d + 6 + 2 - 3d$
  e) $8 + 4e - 3e - 6$
  f) $7f - 1 - 7f + 1$

3 Simplify:
  a) $a + a + b + b + a$
  b) $2c + d + 3c + d$
  c) $5e + 4f - 2e + 3f$
  d) $5g + 2h + 3h - g$
  e) $3j - 2k + 5k - 2j$
  f) $7m - 3n - 4m + 5n$
  g) $2p - 3q + 9p - 4q$
  h) $4s - 7t - s - t$

4 Simplify:
  a) $a^2 + a^2 + a^2$
  b) $b^2 + 6b^2 + 3b^2 + b^2$
  c) $2c^2 + 5c^2 - c^2$
  d) $2d^3 + 2d^3 + 8d^3 - 3d^3$
  e) $e^2 + e + 5e^2 + 4e$
  f) $6f^2 - 4 - f^2 + 15$
  g) $3g^2 + 3g^2 - 2g^2 - 4g^2$
  h) $h^2 - 2h^2 - 3h^2 + 8h^2 - h^2$

5 The width of this rectangle is x cm
  The length is 5 cm more than the width.
  a) Write down an expression for the length of the rectangle.

  x cm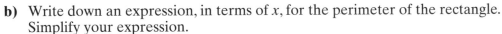

  The perimeter of the rectangle is found using the formula:
    Perimeter = width + length + width + length

  b) Write down an expression, in terms of x, for the perimeter of the rectangle.
    Simplify your expression.

  c) What is the perimeter of the rectangle when $x = 4$?

6 Here are three boxes of books.
  The medium box contains x books.
  The medium box contains 12 more
  books than the small box.
  a) Write down an expression, in terms of x,
    for the number of books in the small box.

    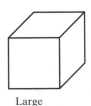
  Small   Medium   Large

  The large box contains twice as many books as the medium box.
  b) Write down an expression, in terms of x, for the number of books in the large box.
  c) Write down an expression, in terms of x, for the **total** number of books in all three
    boxes. Simplify your expression.
  d) What is the total number of books when $x = 20$?

1 Simplify:
  a) $a \times a \times a \times a$ **b)** $b \times b \times b$ **c)** $c \times c \times c \times c \times c$
  d) $7 \times e \times e$ **e)** $8 \times f \times f \times f \times f$ **f)** $4 \times g \times 2 \times g$

2 Simplify:
  a) $x^4 \times x^7$ **b)** $x^3 \times x$ **c)** $y^5 \times y^2$ **d)** $c^{30} \div c^5$
  e) $d^{10} \div d^5$ **f)** $z^5 \times z \times z^3$ **g)** $y^{12} \div y^4$ **h)** $e^3 \times e^3$
  i) $f^6 \div f$ **j)** $g^{20} \div g^5$ **k)** $h^6 \times h^3$ **l)** $k^{10} \times k^2$

3 Simplify:
  a) $7x^5 \times 2x^6$ **b)** $6x^3 \times 3x^3$ **c)** $8x^2 \times x^5$
  d) $5x^5 \times 8x^4$ **e)** $9y \times 4y^7$ **f)** $3x^3 \times 3x^3$
  g) $4x^2 \times 5x^6$ **h)** $3x^5 \times 4x^3 \times 2x$ **i)** $7x^3 \times x^9 \times 6x^5$

4 Simplify:
  a) $15y^8 \div 3y^2$ **b)** $21z^7 \div 7z^3$ **c)** $63x^8 \div 7x^4$
  d) $56z^3 \div 8z$ **e)** $80y^{10} \div 8y^5$ **f)** $16y^6 \div 16y^4$
  g) $13x^7 \div 13x^7$ **h)** $18x^{18} \div 2x^2$ **i)** $15x^5 \div 3x^3$

1 Expand the following brackets:
  a) $3(a + 4)$ **b)** $6(b - 2)$ **c)** $5(2 - c)$
  d) $-9(d + 2)$ **e)** $4(3 + e)$ **f)** $-7(f - 1)$
  g) $-5(3 + g)$ **h)** $2(h - 9)$ **i)** $8(3 - i)$
  j) $-6(j - 3)$ **k)** $-4(2 + k)$ **l)** $7(m + 1)$

2 Expand the following brackets:
  a) $-2(3a + 7)$ **b)** $5(6 - 4b)$ **c)** $3(9c - 5)$
  d) $-4(8 - 5d)$ **e)** $-7(2 + 3e)$ **f)** $6(5 + 3f)$
  g) $10(-g + 4)$ **h)** $-8(6h - 1)$ **i)** $-(7 + 2i)$
  j) $-9(5j - 2)$ **k)** $2(9k + 4)$ **l)** $4(4 - m)$

3 Multiply out the brackets and simplify the results.
  a) $3(x + 7) + 4(x + 1)$ **b)** $8(x + 2) + 5(x - 2)$
  c) $9(x + 1) + 5(x + 2)$ **d)** $6(5x - 2) + 9(2x + 3)$
  e) $8(3x + 2) + 2(4x + 3)$ **f)** $5(3x + 2) + 4(3x - 2)$
  g) $6(x + 6) + 7(3x + 7)$ **h)** $7(x - 1) + 4(2x - 1)$
  i) $4(3x + 2) + 5(2x + 3)$ **j)** $9(8x - 2) + 3(3x + 4)$

4 Multiply out the brackets and simplify the results.
  Take special care when there is a negative number in front of the second bracket.
  a) $5(6x - 1) + 8(2x - 5)$ **b)** $2(6x + 5) - (8x + 9)$
  c) $7(x + 9) + 3(x + 2)$ **d)** $4(4x + 3) - 2(x - 8)$
  e) $3(5x + 7) - 2(3x + 5)$ **f)** $8(x + 2) + 5(x - 7)$
  g) $9(3x - 4) - 9(2x - 7)$ **h)** $6(x - 1) - 4(5x + 2)$
  i) $10(x + 1) + 7(8x - 1)$ **j)** $11x - 6(3x - 7) + 5x$

EXERCISE 7.7

1 Expand:

a) $a(a + 5)$ b) $b(b - 2)$ c) $c(c + 9)$

d) $d(7 + d)$ e) $e(8 - e)$ f) $f(3f - 4)$

g) $g(3 - 2g)$ h) $h(5h + 3)$ i) $i(7 - 6i)$

2 Expand:

a) $7a(a - 2)$ b) $4b(3b + 5)$ c) $6c(3 - c)$

d) $5d(7 + 4d)$ e) $9e(e + 3)$ f) $2f(8 - f)$

g) $8g(7g - 3)$ h) $3h(10h + 1)$ i) $4i(11 - 6i)$

3 Expand and simplify these brackets.
You must show all the steps in your working.

a) $(x + 5)(x + 2)$ b) $(x + 9)(x + 3)$ c) $(x - 6)(x + 4)$

d) $(x + 8)(x - 1)$ e) $(x - 7)(x - 2)$ f) $(x + 6)(x + 5)$

g) $(x - 3)(x + 7)$ h) $(x - 2)(x - 1)$ i) $(x + 5)(x - 7)$

j) $(x + 9)(x + 6)$ k) $(x - 4)(x - 4)$ l) $(x - 3)(x + 2)$

4 A large rectangle is divided into four smaller
rectangles labelled **A**, **B**, **C** and **D**.
The length of one side of the large rectangle
is $(c + 7)$ cm.
The length of the other side of the large
rectangle is $(c + 5)$ cm.

	c	7
c	**A**	**B**
5	**C**	**D**

a) One of the shapes is a square.
   Write down an expression for the area of this square.

b) Write down an expression in terms of c for the area of:
   (i) rectangle **B**   **(ii)** rectangle **C**   **(iii)** rectangle **D**

c) Write down an expression for the total area of the large rectangle.
   Simplify your expression.

5 Expand:

a) $(a + 5)^2$ b) $(b - 6)^2$ c) $(c - 4)^2$ d) $(d + 4)^2$

EXERCISE 7.8

1 Which of the following statements are correct?

a) $2x + 3x \equiv 4x + x$ b) $5 - 9x \equiv -4x$

c) $7x - 3x \equiv x + 3x$ d) $2(x + 3) \equiv 2x + 3$

2 Which of the following expressions are always equal to $4n$?

$6n - 2$	$9n - 5n$	$n \times 4$	$4(n + 1) - 4$	$2n \times 2n$	$n + 3n$

3 a) Show that when $x = 5$, $(x - 2)^2 = x^2 - 4x + 4$

    b) By expanding $(x - 2)^2$ show that you can write $(x - 2)^2 \equiv x^2 - 4x + 4$

EXERCISE 7.9

1 Factorise these expressions:

   a) $5a - 10$ **b)** $12b + 3$ **c)** $21c - 7$

   d) $9d + 63$ **e)** $5 - 40e$ **f)** $6 + 3f$

2 Factorise these expressions:

   a) $6 + 9a$ **b)** $10b - 20$ **c)** $8c + 6$

   d) $20 - 4d$ **e)** $12e - 18$ **f)** $16 + 24f$

3 Factorise these expressions:

   a) $7a + ab$ **b)** $xy - x$ **c)** $5c + cd$

   d) $3xy + 2y$ **e)** $4de + 6e$ **f)** $10e - 15de$

4 Factorise these expressions:

   a) $a^2 + 9a$ **b)** $b^2 - 3b$ **c)** $7c - c^2$

   d) $8d^2 - d$ **e)** $e^2 + 12e$ **f)** $12f - f^2$

   g) $6x^2 + 3xy$ **h)** $21y^2 + 14y$ **i)** $4hk - 2h^2$

EXERCISE 7.10

1 A triangle has sides of length x cm, $3y$ cm and $5z$ cm.
Write down a formula for the perimeter P of the triangle.

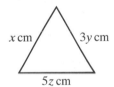

2 A square has length, g cm
Write down a formula in terms of g for:

a) the perimeter P, of the square

b) the area A, of the square.

3 A rectangle has width, x cm
Its length is 10 cm more than its width.

Write down a formula in terms of x for:

   a) the length, L, of the rectangle

   b) the perimeter, P, of the rectangle

   c) the area, A, of the rectangle.

4 Becky buys n books at £p each.
Obtain an equation for the total cost, C, for the books that Becky buys.

5 Pens cost 5p each and pencils cost 8p each.

   a) Nick buys x pens and y pencils.
      Write down an expression in terms of x and y for the total cost of the pens and
      pencils Nick buys.

   b) If the total cost is z pounds, write down an equation in terms of x, y and z

6 The cost of hiring a machine is £20 plus a daily charge of £5 per day.
  a) Find the cost of hiring the machine for 6 days.
  b) Obtain a formula for the cost, £C, of hiring the machine for x days.

7 In a school there are g girls and b boys.
  a) Write an expression for the total number of students in the school.
  b) Write a formula for the total number of students, N, in the school.

8 On Monday Melissa runs at p km/h for q hours.
  Write down a formula for the total distance, D, that Melissa runs on Monday.

EXERCISE 7.11

1 Rearrange these formula to make a the subject:

  a) $b = a - 5$ **b)** $c = 4 + a$ **c)** $e = \dfrac{a}{2}$

  d) $f = 6a$ **e)** $g = 9 - a$ **f)** $h = \dfrac{1}{2}a$

2 Rearrange these formulae to make x the subject:

  a) $y = 3x - 4$ **b)** $y = 5 + 2x$ **c)** $y = \dfrac{x}{7} - 1$

  d) $y = 8 + \dfrac{x}{5}$ **e)** $y = 6x + 1$ **f)** $y = \dfrac{x - 1}{7}$

3 Rearrange these formulae to make w the subject:

  a) $y = wx$ **b)** $y = w + v$ **c)** $y = \dfrac{w}{c}$

  d) $y = 3w - c$ **e)** $y = wh + 2e$ **f)** $y = \dfrac{c}{w}$

4 Rearrange these formulae so that the indicated letter becomes the subject:

  a) $A = lw$ (w) **b)** $P = 2b + 2h$ (h)
  c) $y = 5x - 2$ (x) **d)** $D = s \times t$ (t)
  e) $E = \dfrac{360}{n}$ (n) **f)** $S = 180(n - 2)$ (n)
  g) $C = \pi d$ (d) **h)** $T = x(y - z)$ (x)
  i) $v = u + at$ (t) **j)** $m = \dfrac{y}{x}$ (y)

Chapter 8

Equations

EXERCISE 8.1

Look at the various algebraic statements labelled **A** to **J**:

A	$7x + 2 = 3$	**B**	$6y + 5$
C	πr^2	**D**	$V = L \times B \times H$
E	$5x - 2x = 3x$	**F**	$x^3 = 27$
G	$2c + 3 = c + 3 + c$	**H**	$A = \frac{1}{2}(a + b)h$
I	$x^2 - 9$	**J**	$12x = 30$

1 Which ones are expressions?

2 Which ones are equations?

3 Which ones would you call formulae?

4 Which ones are identities?

EXERCISE 8.2

Do not use your calculator for Questions 1–4.

Show all your working when answering the questions in this exercise.

1 Solve:

    a) $a + 7 = 10$ **b)** $b - 3 = 18$ **c)** $c + 11 = 24$

2 Solve:

    a) $a - 7 = 7$ **b)** $b + 15 = 35$ **c)** $c - 20 = 55$

3 Solve:

    a) $3a = 21$ **b)** $\dfrac{b}{5} = 4$ **c)** $9c = 18$

4 Solve:

    a) $\dfrac{a}{2} = 7$ **b)** $\dfrac{b}{6} = 5$ **c)** $8c = 24$

You can use your calculator for Questions 5 and 6.

5 Solve:

    a) $x + 5 = 27$ **b)** $x - 18 = 7$ **c)** $6x = 54$

    d) $\frac{1}{2}x = 8$ **e)** $x - 13 = 44$ **f)** $\dfrac{x}{4} = 15$

    g) $5x = 80$ **h)** $\dfrac{x}{7} = 21$ **i)** $x + 42 = 42$

6 Solve:

  a) $8t = 48$ **b)** $u - 6 = 4$ **c)** $\frac{1}{4}p = 5$

  d) $20g = 10$ **e)** $5q = 1$ **f)** $w + 8 = 3$

  g) $6y = 5$ **h)** $y + 12 = 5$ **i)** $\frac{p}{6} = -3$

EXERCISE 8.3

Do not use your calculator for Questions 1 and 2.

1 Solve:

  a) $2a + 7 = 15$ **b)** $4b - 1 = 7$ **c)** $9c + 5 = 23$
  d) $5 + 3d = 20$ **e)** $7e - 3 = 4$ **f)** $6f + 13 = 31$
  g) $5g + 3 = 3$ **h)** $1 + 8h = 41$ **i)** $10i - 10 = 10$

2 Solve:

  a) $7 - a = 3$ **b)** $15 - b = 10$ **c)** $20 - c = 14$
  d) $9 - 2d = 5$ **e)** $11 - 3e = 2$ **f)** $8 - 4f = 0$
  g) $25 - 7g = 4$ **h)** $30 - 6h = 0$ **i)** $6 - 2i = 4$

You can use your calculator for Questions 3 and 4.

3 Solve:

  a) $5a + 7 = 17$ **b)** $2b - 1 = 15$ **c)** $3c + 2 = 14$
  d) $9 + 7d = 23$ **e)** $6e - 5 = 25$ **f)** $27 + 4f = 3$
  g) $8g - 3 = -27$ **h)** $9h + 11 = 2$ **i)** $10i - 3 = 7$

4 Solve:

  a) $\dfrac{a + 3}{2} = 4$ **b)** $\dfrac{b - 7}{3} = 5$ **c)** $\dfrac{2c}{5} - 1 = 3$

  d) $\dfrac{9 + d}{4} = 2$ **e)** $\dfrac{3e}{7} + 1 = 4$ **f)** $\dfrac{2 - f}{3} = 1$

  g) $\dfrac{g - 4}{5} = 3$ **h)** $\dfrac{2 - 3h}{7} = 5$ **i)** $12 - \dfrac{i}{8} = 4$

EXERCISE 8.4

Solve:

1 $5(a + 2) = 20$ **2** $6(2b - 1) = 12$ **3** $4(3c + 4) = 16$

4 $2(5 - d) = 6$ **5** $9(4e + 1) = 45$ **6** $3(20 - f) = 21$

7 $7(2g - 3) = 70$ **8** $8(6 + 5h) = 144$ **9** $5(6i + 1) = 35$

10 $3(9 + 4j) = 33$ **11** $6(3k - 7) = 30$ **12** $4(2m - 1) = -20$

EXERCISE 8.5

Solve these algebraic equations, showing the steps of your working clearly.
All the answers are integers, but some may be negative.

1 $3x + 2 = x + 12$ **2** $22 + 5x = 6x$

3 $9x + 4 = 4x + 29$ **4** $x + 5 = 13 - x$

5 $8x + 3 = 48 - x$ **6** $23 - 2x = 33 - 7x$

7 $5x + 3 = 15 - x$ **8** $7 + 4x = 2x + 9$

9 $8x - 5 = x + 9$ **10** $6x = 42 - x$

Expand any brackets and simplify before solving these equations.

11 $6(x - 3) + 2 = 14$ **12** $3(x - 4) = 23 - 2x$

13 $4(x + 10) - 50 - x$ **14** $8(x - 3) - 8$

15 $2(5x + 3) = 7x + 24$ **16** $x + 7 = 5(x - 1)$

17 $4(3x - 5) = 3(2x + 1) + 1$ **18** $5(x - 1) = 3(x - 5)$

19 $7(x + 8) = 4(x + 7) - 14$ **20** $2(5x + 3) = 4(x - 3)$

EXERCISE 8.6

Solve the following quadratic equations:

1 $x^2 = 9$ **2** $x^2 = 49$ **3** $x^2 = 121$

4 $3x^2 = 12$ **5** $5x^2 = 5$ **6** $4x^2 = 36$

7 $x^2 - 1 = 15$ **8** $x^2 + 3 = 28$ **9** $x^2 + 13 = 113$

10 $x^2 - 5 = 76$ **11** $3x^2 + 1 = 13$ **12** $2x^2 - 9 = 9$

13 $\dfrac{x^2}{3} = 48$ **14** $\dfrac{x^2}{5} = 5$ **15** $4x^2 + 3 = 199$

EXERCISE 8.7

Do not use your calculator for Questions 1–6.

1 A CD costs £5. Tim buys x CDs.
  a) Write down an expression, in terms of x, for the total cost of Tim's CDs.
  Tim spends £80 on CDs.
  b) Write down an equation to show this information.
  c) Solve your equation to find how many CDs Tim buys.

2 A rectangle has length $(c + 8)$ cm and width c cm
  a) Write down an expression, in terms of c, for the
  perimeter of the rectangle.

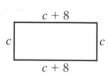

  The perimeter of the rectangle is 60 cm
  b) Write down an equation to show this information.
  c) Solve your equation to find the width and length of the rectangle.

3 Ben earned y euros per hour in January.
In February his earnings were increased by 2 euros per hour.
   a) Write down an expression, in terms of y, for Ben's new earnings per hour.
   b) Write down an expression, using brackets, if Ben works for 10 hours in February.

Ben earns 170 euros in February for 10 hours' work.
   c) Write down an equation for this information.
   d) Solve your equation to find the value of y

4

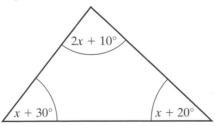

Seyi — I think of a number. I add 7 to my number and then multiply this by 5. The answer is 20

Jamie — I think of a number. I divide my number by 10 and then add 16 to the result. The answer is 36

   a) (i) Use x to stand for Seyi's mystery number.
          Write down an equation for Seyi's mystery number.
      (ii) Solve your equation to find Seyi's number.
   b) Use y to stand for Jamie's mystery number.
      (i) Write down an equation for Jamie's mystery number.
      (ii) Solve your equation to find Jamie's number.

5 An isosceles triangle has two sides of length x cm
   The third side is 4 cm more than each of the other two sides.
   a) Write down an expression for the third side of the triangle.
   b) (i) Write down an expression for the perimeter
          of the triangle.
      (ii) Simplify your expression.

The perimeter of the triangle is 40 cm.
   c) Work out the lengths of all three sides of the triangle.

6 A triangle has angles of $x + 30°, 2x + 10°$ and $x + 20°$

   a) Write down an expression, in terms of x for the sum of the 3 angles.
   b) (i) Use the fact that the angles of a triangle add up to $180°$ to write an equation in
          terms of x
      (ii) Solve your equation to find the value of x
   c) Write down the size of the biggest angle.

You can use your calculator for Questions 7–10.

7 a) Write down an expression for the area of this square.

$t\,\text{cm}$

The area of the square is 225 cm²

b) What is the length of one side of the square?

8 a) (i) Write down an expression, in terms of x for the perimeter of this rectangle.

(ii) Simplify your expression.

$5x - 2$

$2x + 5$
(width)

b) (i) Write down an expression, in terms of x for the perimeter of this triangle.

$x + 12$ $x + 12$

$3x$

(ii) Simplify your expression.

The perimeter of the rectangle is the same as the perimeter of the triangle. All measurements are in cm.

c) (i) Write down an equation to show this information.

(ii) Solve your equation to find the width of the rectangle.

9 a) Write down an expression for this information.
Use n for Marco's mystery number.

I think of a number.
I multiply the number by 7.
I then add 5 to the new total.

Marco

Marco gets the same answer when he adds 29 to his mystery number.

b) Write down an equation for this information.

c) Solve your equation to find Marco's mystery number.

10 Michael and Tanya each think of the same number.
Michael multiplies the number by 4, and then adds 2
Tanya adds 4 to the number, and then multiplies the answer by 2
They both end up with the same answer.

a) Write this information as an equation.

b) Solve your equation, to find the number they both thought of.

EXERCISE 8.8

Use a table like the one in the Example on page 173 to help with these questions.
Use trial and improvement to solve these equations correct to 1 decimal place.

1 The equation $x^2 + 3x = 24$ has a solution between $x = 3$ and $x = 4$

2 The equation $x^2 + x = 33$ has a solution between $x = 5$ and $x = 6$

3 The equation $x^3 + x = 4$ has a solution between $x = 1$ and $x = 2$

4 The equation $x^2 - 5x = 20$ has a solution between $x = 7$ and $x = 8$

5 The equation $x^3 + 7 = 100$ has a solution between $x = 4$ and $x = 5$

6 The equation $x^2 - 5 = 70$ has a solution between $x = 8$ and $x = 9$

7 The equation $2x^2 + x = 2$ has a solution between $x = 0$ and $x = 1$

Use trial and improvement to solve these equations correct to 2 decimal places.

8 The equation $2x^3 - x = 500$ has a solution between $x = 6$ and $x = 7$

9 The equation $x^2(x + 1) = 34$ has a solution between 2 and 3

10 The equation $x^2 + 2x = 7$ has a solution between $x = 1$ and $x = 2$

EXERCISE 8.9

1 Write down the correct inequality sign ($<$ or $>$) between these pairs of numbers:
   a) 6 ... 9
   b) 3.7 ... 2.9
   c) 6....5.97
   d) 3.08 ... 3.8
   e) -5 ... -6
   f) -9 ... 0
   g) -22 ... -20
   h) -6.8 ... -6.9
   i) 20 ... -18

2 Show these inequalities on a number line:
   a) $a \leqslant 2$
   b) $b > -1$
   c) $c \geqslant 5$
   d) $-4 < d < 3$
   e) $-2 \leqslant e \leqslant 3$
   f) $-5 < f \leqslant -1$

3 Write down the possible integer values for n:
   a) $2 < x < 8$
   b) $-3 \leqslant x \leqslant 1$
   c) $1 < x \leqslant 6$
   d) $-5 \leqslant x < 0$
   e) $-8 < x < -3$
   f) $-4 < x \leqslant 2$

4 Solve these inequalities:
   a) $y + 5 \leqslant 9$
   b) $y - 2 \geqslant 3$
   c) $7y > 21$
   d) $5y < 30$
   e) $2y - 3 \geqslant 5$
   f) $7 + 4y \leqslant 27$

5 Solve these inequalities:
   a) $4 \leqslant x + 2 \leqslant 12$
   b) $-2 \leqslant x - 6 < 3$
   c) $-9 < 3x < 24$
   d) $35 < 5x \leqslant 50$
   e) $6 < x - 3 \leqslant 15$
   f) $-8 \leqslant 2x \leqslant -2$

6 Write down the possible integer values for n:
   a) $5 \leqslant n + 1 \leqslant 8$
   b) $4 \leqslant 2n < 16$
   c) $-6 \leqslant 3n \leqslant 12$
   d) $7 < n - 3 < 12$
   e) $-5 < n + 2 \leqslant -2$
   f) $-30 \leqslant 5n < -10$

Chapter 9

Number sequences

EXERCISE 9.1

1 Here are the first three patterns in a sequence of patterns made from sticks:

Pattern 1   Pattern 2   Pattern 3

a) Draw pattern number 4.
b) Copy and complete the table.

Pattern number	1	2	3	4	5
Number of sticks	3	5	7		

c) Explain how you would work out the number of sticks in pattern number 6.
d) Work out how many sticks there are in pattern number 15.

2 The diagram shows how tables are arranged in a hall.
x shows where a chair is placed.

  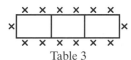

Table 1   Table 2   Table 3

The rest of the tables in the hall follow the same pattern.
a) Draw table number 4.
b) Copy and complete the table.

Table number	1	2	3	4	5
Number of chairs	6	10	14		

c) Explain how you would work out the number of chairs at table number 6.
d) Work out how many chairs at table number 12.

3 Here are the first three patterns in a sequence of patterns made from sticks.

P 1   P 2   P 3

a) Draw P4.
b) Copy and complete the table.

P	1	2	3	4	5
Number of sticks	6	11	16		

c) Work out how many sticks there are in P8.
d) Can a pattern be made from exactly 162 sticks? Explain your answer.

4 Here is a pattern made from squares. Each square has four edges.

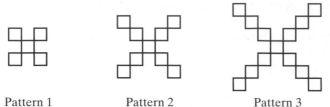

| Pattern 1 | Pattern 2 | Pattern 3 |

a) Draw pattern number 4.
b) Copy and complete the table.

Pattern number	1	2	3	4	5
Number of edges	20	36	52		
Number of squares	5	9	13		

c) Explain how you would work out the number of edges and the number of squares in pattern number 6.
d) Work out how many **(i)** edges, **(ii)** squares there are in pattern number 10.

5 A pattern is made from circles and squares.

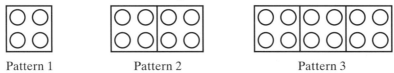

| Pattern 1 | Pattern 2 | Pattern 3 |

a) Draw pattern number 4.
b) Copy and complete the table.

Pattern number	1	2	3	4	5	10	20
Number of squares	1	2	3				
Number of circles	4	8	12				
Total squares + circles	5	10	15				

c) Which pattern number has 60 circles?
d) Which pattern number has a total of 200 squares + circles?

EXERCISE 9.2

1 Write down the first five terms of the number sequences generated by each of these rules.
   a) The first term is 6
      To find the next term add 7 to the previous term.
   b) The first term is 3
      To find the next term multiply the previous term by 4.
   c) The first term is 28
      To find the next term subtract 4 from the previous term.

d) The first term is 486
To find the next term divide the previous term by 3
e) The first term is 7
To find the next term subtract 1 from the previous term.
f) The first term is 3
To find the next term double the previous term.
g) The first term is 7
To find the next term, double the previous term and add 6

2 For each sequence: **(i)** write down the next two terms
(ii) explain the rule in words.

a) 4, 6, 8, 10, 12 …   **b)** 5, 10, 15, 20, 25 …   **c)** 24, 21, 18, 15, 12 …
d) 70, 63, 56, 49, 42 …   **e)** 6, 10, 14, 18 …   **f)** 15, 9, 3, –3, –9 …

3 Find the missing terms in these number sequences.

a) 3, 7, ?, 15, ?, ?   **b)** ?, 5, 8, 11, ?, ?
c) 15, ?, ?, 21, 23, 25, ?   **d)** 39, ?, ?, 30, 27, 24, ?
e) ?, ?, ?, 18, 23, 28, 33   **f)** 9, ?, 5, ?, ?, –1, –3, –5

4 **(i)** Write down the next term in the following sequences.
(ii) Describe how you found the next term.

a) 1, 8, 27, 64, …   **b)** 1, 3, 9, 27, 81, …
c) 2, 2, 4, 6, 10, 16, …   **d)** 49, 36, 25, 16, 9, …
e) 3, 5, 8, 12, 17, 23 …   **f)** 64, 32, 16, 8 …

5 Look at this number sequence: 3, 7, 11, 15, 19, …
a) Write down the next three terms in the sequence.
b) Explain how you found your answer.

The 50th term of the sequence is 199
c) Write down the 51st term.
d) Will 500 be a term in this sequence?
Explain your answer.

6 Andy has been doing an investigational GCSE coursework task.
He gets this sequence of numbers:

12, 17, 22, 27, 32, …

a) Write down the next three terms in Andy's number sequence.
b) Write a rule for Andy's number sequence.

7 Emily has been working with even numbers.
Emily is not quite right.
Rewrite Emily's statement
so that it is correct.

The rule for odd numbers is: 'add 2 to the previous term

Emily

9 Number sequences   **41**

8 Look at this number sequence: $2, 5, 8, 11, 14, \ldots$
  a) Write down the next three terms in the sequence.
  b) Explain how you found your answer.

  The 50th term of the sequence is 149
  c) Write down the 51st term.
  d) Will 99 be a term in this sequence? Explain your answer.

9 Look at this number sequence: $7, 13, 19, 25, 31, \ldots$
  a) Write down the next three terms in the sequence.
  b) Explain how you found your answer.

  The 100th term of the sequence is 601
  c) Write down the 101st term.
  d) Will 100 be a term in this sequence? Explain your answer.

EXERCISE 9.3

1 Write down the first five terms of the following sequences:
  a) nth term $= 5n$ **b)** nth term $= 6n$ **c)** nth term $= n - 1$
  d) nth term $= n + 2$ **e)** nth term $= n + 4$ **f)** nth term $= n - 5$
  g) nth term $= 2n + 1$ **h)** nth term $= 3n - 1$

2 The nth term of a number sequence is given by the formula $4n - 3$
  a) Write down the first five terms of the number sequence.
  b) Work out the 20th term.

3 The nth term of a number sequence is given by the formula $\frac{1}{2}(10n + 3)$
  a) Write down the first five terms of the number sequence.
  b) Work out the 10th term.

4 Write down the first five terms of the following sequences:
  a) nth term $= n^3 - 1$
  b) nth term $= n^2 + 1$
  c) nth term $= 2n^2$
  d) nth term $= n^3 + n^2$

> Remember:
> n^2 means $n \times n$
> n^3 means $n \times n \times n$

5 The nth term of a number sequence is given by the formula $\dfrac{n^2 + 2}{2}$
  a) Write down the first five terms of the number sequence.
  b) Work out the 30th term.

EXERCISE 9.4

1 Find an expression for the nth term of the following sequences:
  a) $3, 6, 9, 12, 15, \ldots$ **b)** $5, 10, 15, 20, 25, \ldots$
  c) $8, 16, 24, 32, 40, \ldots$ **d)** $11, 22, 33, 44, 55, \ldots$
  e) $100, 200, 300, 400, 500, \ldots$ **f)** $20, 40, 60, 80, 100, \ldots$

2 Find an expression for the nth term of the following sequences:
  a) $5, 8, 11, 14, 17, \ldots$ **b)** $4, 9, 14, 19, 24, \ldots$
  c) $3, 7, 11, 15, 19, \ldots$ **d)** $5, 7, 9, 11, 13, \ldots$
  e) $13, 23, 33, 43, 53, \ldots$ **f)** $5, 12, 19, 26, 33, \ldots$

3 Find an expression for the nth term of the following sequences:
  a) $9, 7, 5, 3, 1, \ldots$
  c) $90, 80, 70, 60, 50, \ldots$
  b) $20, 16, 12, 8, 4, \ldots$
  d) $23, 20, 17, 14, 11, \ldots$

4 The first five terms in an arithmetic sequence are:

  $8, 14, 20, 26, 32, \ldots$

  a) Find the value of the tenth term.
  b) Write a formula for the nth term.

5 The first four terms in an arithmetic sequence are:

  $43, 38, 33, 28, 23, \ldots$

  a) Find the value of the first negative term.
  b) Write a formula for the nth term.

6 Robert has been making patterns with sticks.
  Here are his first three patterns.
  a) Draw pattern 4
  b) Work out the number of sticks in pattern 5
  c) Give a formula for the number of sticks, s, in pattern n
  d) Explain how your formula relates to the way the sticks fit together.

Pattern 1   Pattern 2   Pattern 3

7 The seventh term of an arithmetic sequence is 34 and the eighth term is 40
  a) Write down value of the common difference for this sequence.
  b) Work out the value of the first term.
  c) Find an expression for the nth term of the sequence.
      Check that your expression works when $n = 5$ and $n = 6$

EXERCISE 9.5

Do not use your calculator for Questions 1–3.

1 Look at this pattern of trapeziums made from sticks.

Pattern 1   Pattern 2   Pattern 3

  a) Draw the next pattern.
  b) Copy and complete the table.
  c) How many sticks would there be in pattern 7?
  d) Find a formula for the number of sticks, S, in pattern n

Pattern number	1	2	3	4	5
Number of sticks	5	9	13		

2 (i) Write down the next two terms in each of these number sequences.
  (ii) Find the 10th term.
  (iii) Give a formula for the nth term in each case.
  a) $0, 3, 8, 15, 24, \ldots$
  c) $8, 11, 16, 23, 32, \ldots$
  e) $201, 204, 209, 216, 225, \ldots$
  b) $3, 6, 11, 18, 27, \ldots$
  d) $11, 14, 19, 26, 35, \ldots$
  f) $-4, -1, 4, 11, 20, \ldots$

3 Here are the first three patterns in a sequence of patterns made from hexagons.

Pattern 1   Pattern 2   Pattern 3

a) Draw pattern number 4.

Pattern number	1	2	3	4	5
Number of black hexagons	2	3	4		
Number of white hexagons	2	4	6		
Total number of hexagons	4	7	10		

b) Copy and complete the table.
c) Work out the **total** number of hexagons in pattern number 10
d) Find a formula for the **total** number of hexagons, H, in pattern number n

4 To work out a rule for the nth term in this pattern, we can put two trapeziums together to make rectangles like this:

 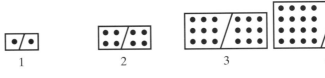 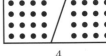

  1   2   3   4

No. of dots in rectangle: $1 \times 2 = 2$ $2 \times 4 = 8$ $3 \times 6 = ?$ $? \times ? = ?$

No. of dots in trapezium: $\dfrac{1 \times 2}{2} = 1$ $\dfrac{2 \times 4}{2} = 4$ $\dfrac{3 \times 6}{2} = ?$ $\dfrac{? \times ?}{2} = ?$

a) How many dots are there in rectangle:
  (i) 3   **(ii)** 4   **(iii)** 5?
b) How many dots are there in trapezium:
  (i) 3   **(ii)** 4   **(iii)** 5?
c) Carrying on the pattern:
  (i) the 10th rectangle would have $10 \times (10 + 10) = ?$ dots

  (ii) the 10th trapezium would have $\dfrac{10 \times (? + ?)}{2} = ?$ dots

  (iii) the nth rectangle would have $? \times (? + ?)$ dots

  (iv) the nth trapezium would have $\dfrac{? \times (? + ?)}{?}$ dots.

Fill in the missing values.

Chapter 10

Coordinates and graphs

EXERCISE 10.1

1 Write down the coordinates of the points *A* to *H*.

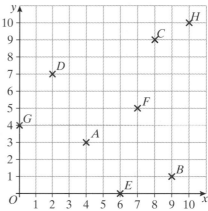

2 Write down the coordinates of the points *A* to *H*.

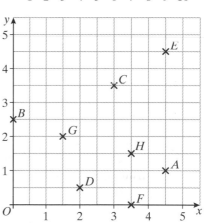

3 **a)** Write down the coordinates of:
   (i) *A*   **(ii)** *B*.

   b) Write down the coordinates of the midpoint of the line segment *AB*.

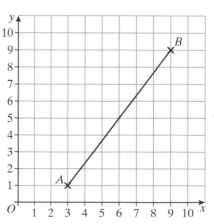

4 Draw a pair of axes on a grid and label both axes from 0 to 10
  a) Plot the points $A(7, 1)$ and $B(1, 9)$
    Join the points with a straight line.
  b) Mark the midpoint, M, of the line segment AB.
  c) Write down the coordinates of M.

5 Draw a pair of axes on a grid and label both axes from 0 to 10
  a) Plot the points $A(2, 1)$, $B(6, 1)$, $C(6, 5)$, and $D(2, 5)$
  b) Join the points in order, to form a closed shape.
  c) What is the name of the shape?

  Mark the centre, M, of your shape.
  d) Write down the coordinates of M.

6 Draw a pair of axes on a grid and label both axes from 0 to 6
  a) Plot the points $A(1, 2)$, $B(6, 2)$ and $C(6, 5)$. Join A to B and B to C.

  A fourth point, D, is needed to form a rectangle.
  b) Write down the coordinates of D.

7 The point A has coordinates $(9, 4)$
  The point B has coordinates $(3, 6)$
  M is the midpoint of the line segment AB.
  Find the coordinates of M.

8 The point A has coordinates $(14, 3)$
  The point B has coordinates $(4, 11)$
  M is the midpoint of the line segment AB.
  Find the coordinates of M.

EXERCISE 10.2

1 Using the diagram, write down the
  coordinates of A, B, C, D and E.

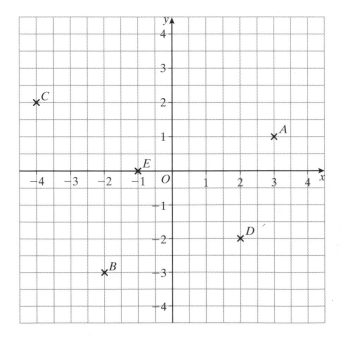

2 a) Which point is at $(-4, -2)$?
b) What are the coordinates of L?
c) Which point is midway between $(2, -6)$ and $(8, 0)$?
d) What are the coordinates of H?
e) Which point has the **largest** y coordinate?
f) Which point has the **smallest** x coordinate?
g) Which point has the **same** x coordinate and y coordinate?

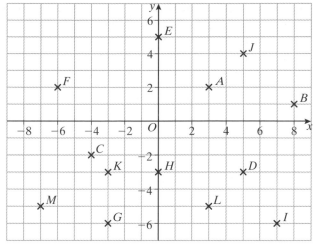

3 Draw a pair of axes on a grid and label both axes from -4 to 4
a) Plot the points $A(-1, 4)$ and $B(3, -2)$
b) Join the points with a straight line.
Mark the midpoint, M, of the line segment AB.
c) Write down the coordinates of M.

4 Draw a pair of axes on a grid and label both axes from -4 to 5
a) Plot the points $A(5, 1)$, $B(5, -3)$, $C(-1, -3)$, and $D(-1, 1)$
b) Join the points in order, to form a closed shape.
c) What is the name of the shape?
d) Mark the centre, M, of your shape.
Write down the coordinates of M.

5 Draw a pair of axes on a grid and label both axes from -6 to 6
a) Plot the points $A(3, 2)$, $B(-1, 2)$ and $C(-3, -2)$. Join A to B and B to C.
b) A fourth point, D, is needed to form a parallelogram.
Write down the coordinates of D.
c) Mark the centre, M, of your shape.
Write down the coordinates of M.

6 The point A has coordinates $(8, -2)$. The point B has coordinates $(-4, -6)$
M is the midpoint of the line segment AB. Find the coordinates of M.

7 The point A has coordinates $(-8, 0)$. The point B has coordinates $(4, -6)$
M is the midpoint of the line segment AB. Find the coordinates of M.

8 Follow these instructions carefully.
Draw a coordinate grid and label both axes from -10 to 10 on the x axis and -4 to 4 on the y axis. Now draw line segments as follows:

From $(-8, 3)$ to $(-8, -3)$	From $(0, 3)$ to $(4, 3)$
From $(-8, 0)$ to $(-5, 3)$	From $(6, 3)$ to $(6, -3)$
From $(-8, 0)$ to $(-5, -3)$	From $(6, 3)$ to $(9, 3)$
From $(-2, 3)$ to $(-2, -3)$	From $(6, 0)$ to $(8, 0)$
From $(2, 3)$ to $(2, -3)$	From $(6, -3)$ to $(9, -3)$

You should find that you have named a 4-sided shape. Draw a sketch of this shape.

EXERCISE 10.3

Write down the coordinates of the **vertices** (corners) on each of these cuboids.

1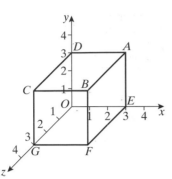

2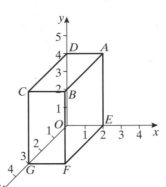

3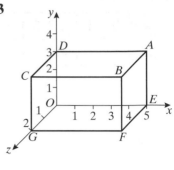

4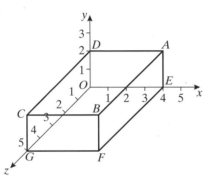

5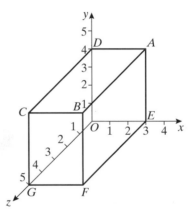

EXERCISE 10.4

1 Write down the equation of each of the lines labelled A to E.

2 Draw a pair of axes on a grid and label both axes from −6 to 6
Draw these lines and label them with their letter:

a) $x = 3$ **b)** $y = 2$ **c)** $x = -4$
d) $y = -5$ **e)** $y = x$ **f)** $x = -y$

3 Write down an equation for the **a)** x axis and **b)** y axis.

For Questions **4** to **10**, complete the table of values for the given equation and draw the graph. A rough set of axes next to each question shows you how to draw your axes.

4 $y = x + 1$

x	0	1	2	3	4
y	1		3		5

5 $y = 2x + 1$

x	0	1	2	3	4
y	1	3			9

6 $y = 4x$

x	0	1	2	3	4
y	0	4		12	

7 $y = 3x - 2$

x	0	1	2	3	4
y	-2	1			10

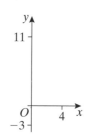

8 $y = 2x - 1$

x	-2	-1	0	1	2
y		3			3

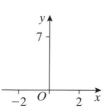

9 $y = 4 - x$

x	-2	-1	0	1	2
y		5		3	

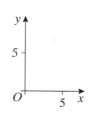

10 $x + y = 5$

x	0	1	2	3	4	5
y			3	2		

11 a) Draw a pair of axes so that:
   the x values start at 0 and end at 5
   the y values start at 0 and end at 10
   b) Draw the graph of $y = x + 4$
   c) Use your graph to find:
      (i) the value of y when $x = 2.3$
      (ii) the value of x when $y = 7.8$

12 a) Draw a pair of axes so that:
   the x values start at 0 and end at 5
   the y values start at 0 and end at 18
   b) Draw the graph of $y = 3x + 1$
   c) Use your graph to find:
      (i) the value of y when $x = 2.7$
      (ii) the value of x when $y = 12.5$

10 Coordinates and graphs   49

13 a) Draw a pair of axes so that:
the x values start at -4 and finish at 4
the y values start at -4 and finish at 8
b) Draw the graph of $x + y = 2$
c) Use your graph to find:
    (i) the value of y when $x = 3.5$
    (ii) the value of x when $y = 2.8$

14 Draw the following graphs:
a) $y = x + 6$ for $0 \leqslant x \leqslant 4$
b) $y = 2x$ for $0 \leqslant x \leqslant 4$
c) $y = 4x - 1$ for $0 \leqslant x \leqslant 4$

EXERCISE 10.5

Find the gradient of the lines marked in Questions **1–8** below.

1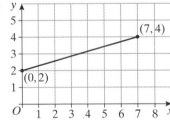

2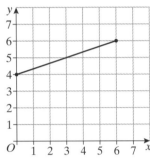

3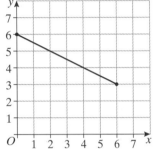

4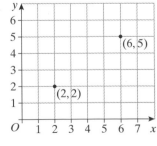

5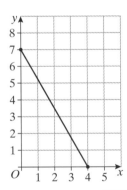

6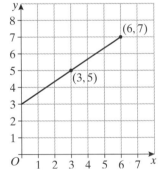

7

8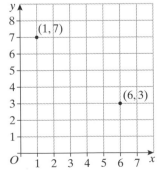

EXERCISE 10.6

1 Write down the gradients of the lines:

a) $y = 6x$

b) $y = \frac{1}{3}x$

c) $y = 2x - 7$

d) $y = -3x$

e) $y = 8 - 5x$

f) $y = x + 2$

2 The diagram shows the graph corresponding to a linear function of x
   a) Write down the coordinates of the points P and Q on the line.
   b) Write down the gradient and the y intercept of the line.
   c) Hence find the equation of the straight line.

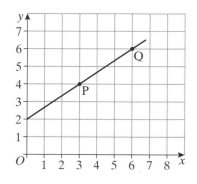

3 The diagram shows the graph of a linear function of x

   a) Find the gradient and the y intercept of the line.
   b) Hence write down the equation of the straight line.

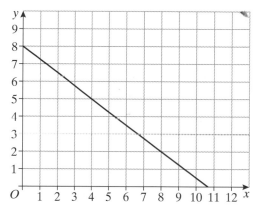

EXERCISE 10.7

1 a) On the same axes draw the graphs of:
   $y = 3x - 1$ and $y = 2x$
   b) Write down the coordinates of the point of intersection.
   c) Hence solve the simultaneous equations:
   $y = 3x - 1$
   $y = 2x$

2 a) On the same axes draw the graphs of:
   $y = 4$ and $y = 4x$
   b) Write down the coordinates of the point of intersection.
   c) Hence solve the simultaneous equations:
   $y = 4$
   $y = 4x$

3 a) On the same axes draw the graphs of:

$y = 3x$ and $y = x + 2$

b) Write down the coordinates of the point of intersection.

c) Hence solve the simultaneous equations:

$y = 3x$
$y = x + 2$

4 a) On the same axes draw the graphs of:

$y = x + 1$ and $x + y = 7$

b) Write down the coordinates of the point of intersection.

c) Hence solve the simultaneous equations:

$y = x + 1$
$x + y = 7$

5 a) On the same axes draw the graphs of:

$y = x - 3$ and $x = 4$

b) Write down the coordinates of the point of intersection.

c) Hence solve the simultaneous equations:

$y = x - 3$
$x = 4$

EXERCISE 10.8

1 a) Complete a copy of the table of values for $y = x^2 + 1$

x	-3	-2	-1	0	1	2	3
x^2	9		1			4	
$+1$	$+1$		$+1$			$+1$	
y	10		2			5	

b) Draw the graph of $y = x^2 + 1$

2 a) Complete a copy of the table of values for $y = x^2 - 2$

x	-4	-3	-2	-1	0	1	2	3	4
x^2	16			1	0	1		9	
-2	-2			-2		-2		-2	
y	14			-1				7	

b) Draw the graph of $y = x^2 - 2$

c) Use your graph to solve the equation $x^2 - 2 = 0$

3 a) Complete a copy of the table of values for $y = 2x^2$

x	-4	-3	-2	-1	0	1	2	3	4
x^2		9			0		4		
y		18					8		

b) Draw the graph of $y = 2x^2$

4 a) Complete a copy of the table of values for $y = 2x^2 + 3$

x	-4	-3	-2	-1	0	1	2	3	4
x^2		9					4		
$2x^2$		18					8		
$+3$		$+3$					$+3$		
y		21					11		

b) Draw the graph of $y = 2x^2 + 3$

5 a) Complete a copy of the table of values for $y = x^2 - x - 5$

x	-4	-3	-2	-1	0	1	2	3	4
x^2	16	9						9	
$-x$	$+4$	$+3$						-3	
-5	-5	-5						-5	
y	15	7						1	

b) Draw the graph of $y = x^2 - x - 5$
c) Use your graph to solve the equation $x^2 - x + 5 = 0$

6 Draw the following graphs for $-3 \leqslant x \leqslant 3$:
a) $y = 3x^2$ **b)** $y = x^2 - 5$ **c)** $y = 2x^2 - 2$
d) $y = x^2 - x$ **e)** $y = 2x^2 - 2x$ **f)** $y = x^2 + 2x - 2$

Chapter 11

Measurements

EXERCISE 11.1

Do not use your calculator for Questions 1–4.

1 Change these times to the 24 hour clock times.

a)
am

b)
pm

c)
pm

d)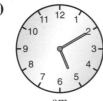
am

e)
am

f)
pm

2 These digital clocks are showing 24 hour clock times.

| 15:18 | 07:28 | 20:20 | 16:51 | 03:49 | 10:09 |

Change them to 12 hour clock times.
Don't forget to write 'am' or 'pm'.

3 The time is 9.52 am.
George had to wait another 4 hours and 23 minutes for his favourite TV programme.
What time is George's TV programme?

4 Peter finishes school at 15 40.
It is now 13 52.
How much longer must Peter wait before school finishes?

You can use your calculator for Questions 5–9.

5 Change these hours to minutes:
   a) 3 hours   b) 1.5 hours   c) $\frac{1}{4}$ of an hour
   d) 0.75 hours   e) 2.3 hours   f) 5.8 hours

6 Change these minutes to hours:
   a) 240 minutes   b) 600 minutes   c) 130 minutes
   d) 90 minutes   e) 20 minutes   f) 135 minutes

7 Look at this digital clock. | 19:48 |

What should the display show in:
a) $\frac{1}{4}$ of an hour's time? **b)** $2\frac{1}{2}$ hours' time? **c)** $1\frac{3}{4}$ hours' time?

8 Here is part of a bus timetable.

Esther catches the 10 31 bus from Sparrow Lane.
a) At what time should the bus arrive at Robin Crescent?

Chris arrives at the High Street at 9.59 am
b) **(i)** How long should Chris expect to wait for a bus to take him to Hawk Road?
(ii) How long should Chris expect the bus journey to take?

High Street	09 23	09 53	10 26	10 50
Sparrow Lane	09 28	09 58	10 31	–
Raven Avenue	09 37	–	–	–
Seagull End	09 45	–	10 46	11 02
Hawk Road	09 52	–	–	11 09
Robin Crescent	10 02	10 28	11 00	11 19

9 Here is part of a train timetable.

Carol plans to catch the 20 48 from Highville to go to Lowden.
a) How long should Carol expect the journey to take?

Carol doesn't arrive at Highville station until 20 56
b) What time should Carol expect to arrive in Lowden?

Highville	19 52	20 14	20 48	21 06
Smalltown	20 12	–	21 08	21 26
Largeford	20 17	20 35	–	–
Littlewood	20 32	20 40	–	21 42
Bighampton	20 48	–	–	21 58
Lowden	21 02	21 08	21 36	22 13

EXERCISE 11.2

Do not use your calculator for Questions 1–7.

1 Write down a sensible metric unit that should be used to measure:
a) the length of a shoe
b) the distance between London and Brighton
c) the length of a pencil
d) the weight of a pen
e) the height of a block of flats
f) the capacity of a teaspoon
g) the weight of your friend
h) the length of an ant
i) the amount of wine in a large barrel
j) the weight of an elephant
k) the length of a classroom
l) the capacity of a cup

2 Convert the following measurements to metres:
a) 3546 mm **b)** 423 cm **c)** 6 km **d)** 85 cm **e)** 0.7 km
f) 135 cm **g)** 7.1 km **h)** 23 km **i)** 4 cm

3 Change the following measurements to centimetres:
a) 7 m **b)** 23 mm **c)** 7 mm **d)** 8.3 m **e)** 92 mm
f) 769 mm **g)** 16.3 mm **h)** 6.25 m **i)** 0.578 m

4 Change the following measurements:
a) 2.9 tonnes to kilograms
b) 20 kilograms to tonnes
c) 0.26 kilograms to grams
d) 7.2 kilograms to grams
e) 8.05 kilograms to grams
f) 800 grams to kilograms

5 Change the following measurements:
a) 7 litres to millilitres
b) 8904 millilitres to litres
c) 520 millilitres to centilitres
d) 8.12 litres to centilitres
e) 0.9 litres to millilitres
f) 64 centilitres to litres

6 Convert the following measurements:
a) 200 metres to kilometres
b) 9000 metres to kilometres
c) 34 centimetres to millimetres
d) 4701 metres to kilometres
e) 76 centimetres to millimetres
f) 5 metres to millimetres

7 Write down sensible estimates using appropriate metric units for:
a) the length and width of your bedroom
b) the height of bedroom door
c) the height of a popstar
d) the weight of a popstar
e) the capacity of a bottle of shampoo
f) the weight of an egg
g) the length of your toothbrush
h) the weight of 10 apples

8 Change the following measurements:
a) 8100 cm^2 to m^2
b) 0.92 m^3 to cm^3
c) 3 m^3 to cm^3
d) 1.2 cm^3 to mm^3
e) 100 mm^3 to cm^3
f) 5 m^2 to cm^2
g) 0.07 m^3 to cm^3
h) 0.91 cm^3 to mm^3
i) 4.49 cm^3 to mm^3
j) 2.8 cm^2 to mm^2
k) 800 mm^2 to cm^2
l) 6.3 m^3 to cm^3
m) $57\,000 \text{ mm}^3$ to cm^3
n) 780 mm^2 to cm^2

EXERCISE 11.3

Do not use your calculator for Questions 1–3.

1 Write down a sensible imperial unit that should be used to measure:
a) the height of a two storey house
b) the length of a garden
c) the weight of a baby sparrow
d) the distance between Glasgow and London
e) the length of a pen
f) the volume of milk in a large carton
g) the weight of a sack of potatoes

2 Change:
a) 150 miles to kilometres
b) 500 miles to kilometres
c) 320 kilometres to miles
d) 30 miles to kilometres
e) 24 kilometres to miles
f) 40 kilometres to miles

3 Change:
a) 20 centimetres to inches
b) 4 litres to pints
c) 18 litres to gallons
d) 120 centimetres to feet
e) 40 gallons to litres
f) 36 feet to metres

You can use your calculator for Questions 4–7.

4 Change:
   a) 55 miles to kilometres   **b)** 18 gallons to litres
   c) 42 pints to litres   **d)** 24 inches to centimetres
   e) 80 miles to kilometres   **f)** 35 kilograms to pounds

5 Jason is 6 feet 4 inches tall.
   a) Change:
        (i) 6 feet to metres   **(ii)** 4 inches to centimetres
   b) What is Jason's height in metres?
   Jason weighs 160 lb
   c) What is Jason's weight in kilograms?
        Give your answer correct to 2 significant figures.

6 Steve buys 10 gallons of beer for a party.
   Approximately how many litres of beer does Steve buy?

7 Johnny drives 370 miles.
   What is 370 miles in kilometres?

EXERCISE 11.4

Do not use your calculator for Questions 1–7.

1 Lisa cycles 40 miles in 5 hours.
   What is her average speed?

2 A car travels 50 miles in one hour 15 minutes.
   What is the average speed of the car in miles per hour?

3 David cycles for 2 hours at an average speed of 10 km/h
   How far does David cycle?

4 Martin drives 200 km at an average speed of 40 km/h
   How long does the journey take?

5 A car travels 3 hours at an average speed of 120 km/h
   How far does the car travel?

6 A car travels for 8 hours at an average speed of 50 mph
   How far does the car travel?

7 Michelle jogs for 30 minutes at an average speed of 7 km/h
   How far does Michelle jog?

You can use your calculator for Questions 8–11.

8 Ian jogs for 40 minutes at an average speed of 80 metres per minute.
   How far does Ian run?

9 Debbie drives 120 km at an average speed of 100 km/h.
   How long in minutes does the journey take?

10 Robert runs 100 metres in 20 seconds.
   a) What is his average speed in metres per second?
   b) What is Robert's average speed in kilometres per hour?

11 An intercity train travels at 160 km/h
   a) How far does it travel in:
       (i) 30 minutes   **(ii)** 45 minutes   **(iii)** 1 hour 15 minutes?
   b) How long in minutes does it take to travel:
       (i) 80 km   **(ii)** 8 km   **(iii)** 200 km?
   c) What is the speed of the train in metres per second?
       Give your answer correct to 1 decimal place.

EXERCISE 11.5

1 The density of water is 1000 kg/m^3
   a) Find the mass of:
       (i) 30 m^3 of water **(ii)** 0.4 m^3 of water **(iii)** 0.6 m^3 of water **(iv)** 8 m^3 of water
   b) What is the volume of:
       (i) 500 kg of water **(ii)** 750 kg of water **(iii)** 2500 kg of water **(iv)** 4000 kg of water?

You can use your calculator for Questions 2 and 3.

2 An aluminium hubcap has a volume of 2620 cm^3
   The mass of the hubcap is 720 g
   Give all your answers to 2 significant figures.
   a) Work out the density of aluminium in g/cm^3
   b) What is the mass in grams of 1 cm^3 of aluminium?
   c) What is the mass in grams of 1 m^3 of aluminium?
   d) What is the mass in kilograms of 1 m^3 of aluminium?
   e) What is the density of aluminium in kg/m^3?

3 The density of silver is 10.49 g/cm^3
   Give all your answers to 3 significant figures.
   a) Find the volume of silver used to make a silver ring of mass 35 g
   b) Find the volume of silver used to make a silver spoon of mass 200 g

   A silver coin has a volume of 3 cm^3
   c) What is the mass of the coin?

EXERCISE 11.6

1 Write down the   **(i)** lower bound   **(ii)** upper bound of these numbers, which have all
   been rounded to the nearest whole number:
   a) 685   **b)** 3   **c)** 37   **d)** 70   **e)** 120

2 Write down the   **(i)** lower bound   **(ii)** upper bound of these measurements:
   a) A line that is 6 cm long to the nearest centimetre.
   b) A house that is 12 m long to the nearest metre.
   c) An orange weighing 132 g to the nearest gram.
   d) A woman who is 135 cm tall to the nearest centimetre.
   e) The distance between two towns that is 156 km to the nearest kilometre.
   f) A child weighing 46 kg to the nearest kilogram.

CHAPTER 12

Interpreting graphs

1 This conversion graph converts between kilometres and miles.

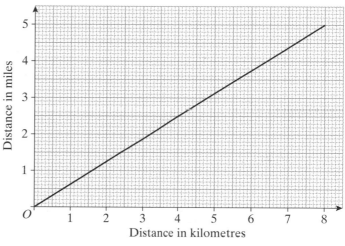

a) Harry travels 6 kilometres. What is this distance in miles?

b) Kate travels 2 miles. What is this distance in kilometres?

c) The distance between the church and the school is 4.3 kilometres. What is this distance in miles?

d) Rick walks 4.3 miles. How far does he walk in kilometres?

e) Paul jogs 5 kilometres. How far is this in miles?

f) Mary wants to convert 0.8 miles to kilometres. What is 0.8 miles in kilometres?

2 This conversion graph converts between pounds (lb) and kilograms (kg).

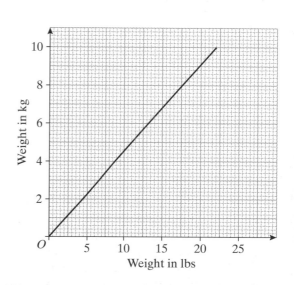

a) A new-born baby weighs 8 lbs What is the baby's weight in kg?

b) A bag of potatoes weighs 5 kg What is the weight of the bag of potatoes in lbs?

c) Barbara buys some chocolate for her bakery. The chocolate in Box A weighs 18 lbs The chocolate in Box B weighs 8.4 kg Both boxes cost the same. Which is the better buy? Give reasons for your answer.

3 The graph shows Jean's weight during the first 12 weeks of her diet.

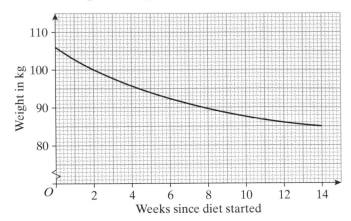

a) How much did Jean weigh at the start of her diet?
b) How much did Jean weigh after: **(i)** 2 weeks **(ii)** 5 weeks?
c) How long did it take for Jean to lose 20 kg?

4 This conversion graph converts the temperature in Celsius to the temperature in Fahrenheit.

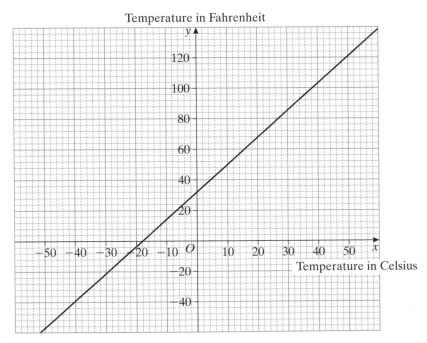

a) Use the graph to convert 0 °C to °F
b) Diana went to Spain. The temperature when she arrived was 28 °C
What is 28 °C in degrees Fahrenheit?

c) Nick went to Siberia. The temperature when he arrived was $-17\,°F$
What is $-17\,°F$ in degrees Celsius?

d) The temperature of a liquid fell from $40\,°C$ to $20\,°C$
What was the total fall in temperature in degrees Fahrenheit?

5 a) Use the fact that 10 litres $\approx$ 17.5 pints to produce a conversion graph between litres and pints.

b) Use your graph to convert:
    (i) 10 pints   **(ii)** 6 pints to litres.

c) Use your graph to convert
    (i) 7 litres   **(ii)** 2 litres to pints.

6 Water runs out of a hole in the bottom of a container.
The water runs out at a steady rate.
The diagram below shows how the depth of water in the container varies over time.

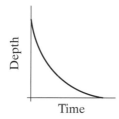   

    A   **B**   **C**

Say which container, A, B or C, best matches this graph.
Explain your reasoning.

EXERCISE 12.2

Do not use your calculator for Questions 1–3.

1 Uzma cycles to her friend.
Here is a description of her journey:

- Uzma leaves her house at 8 am
- She cycles 10 km to arrive at her friend's house at 9.30 am
- Uzma spends 75 minutes with her friend.
- Uzma then cycles back towards her home but stops after 4 km, at 11.15 am to have a rest.
- She rests for $\frac{1}{4}$ hour.
- She then cycles straight home, arriving at 12.30 pm

a) Draw a travel graph of Uzma's journey.
b) What is Uzma's average speed, in km per hour, on her way to her friend?

2 Every morning Ivan walks to
school and walks home after school.
The travel graph shows his journey to school
and his stay at school one Friday.
   a) What time does Ivan arrive at his
        school on Friday?
   b) How long does Ivan stay at school
        on Friday?

After school, Ivan walks slowly home at a
constant speed. But on Friday he stops after
half an hour to visit his friend who lives
1 km from school.
He stays at his friend's house for an hour and
then continues walking home at a constant
speed.
He arrives home at 6 pm
   c) Copy and complete the travel graph for Ivan's journey.
   d) What is Ivan's average speed, in km per hour, on the way to school?

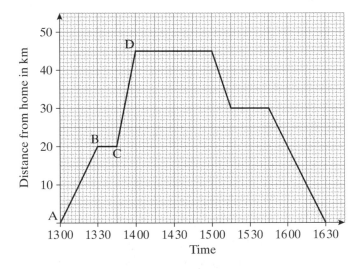

3 One Saturday Maya drives from Liverpool to Manchester to visit her grandmother. She
stops along the way to get some flowers.
On her way home she stops for petrol and a coffee before returning to Liverpool.
The travel graph shows her journey.

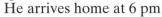

   a) How long does Maya stop for flowers?
   b) What time does Maya get to her grandmother?
   c) How far is her grandmother from Maya's house?
   d) How long did Maya stop for petrol and coffee?
   e) Is Maya's speed greater between C and D or between C and D?
        Explain how you can tell.
   f) What time does Maya arrive home?

You can use your calculator for Questions 4–9.

4 Claire leaves her house for school at 7 am each morning.
   She walks to the bus stop and waits for her bus.
   Here is a distance–time graph for Claire's journey to school.

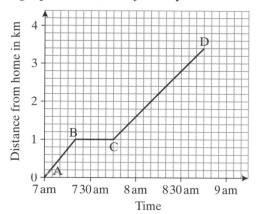

a) What time does Claire reach the bus stop?
b) How long does she have to wait for the bus?
c) How far away from Claire's house is: **(i)** the bus stop   **(ii)** the school?
d) What is Claire's average speed, in km per hour, on her journey:
    (i) from A to B   **(ii)** from C to D?

5 One morning Jamie leaves Crantown to visit his girlfriend in Daly.
   That same day Peter leaves Daly to visit his Aunt in Crantown.
   Here is a travel graph of their journeys.

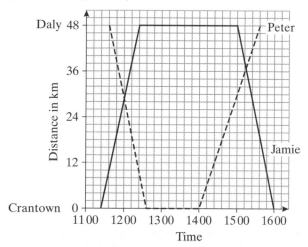

a) What is the distance between Crantown and Daly?
b) What time does Peter leave Daly?
c) What time does Jamie leave Crantown?
d) How far away from Crantown are they when they meet for the first time?
e) How long does Peter spend in Crantown?
f) What time do they meet on their journey home?

6 Geoff went by coach to Bradingham Gardens for lunch.
The coach went directly to the gardens from the coach stop.
On the way home the coach stopped at a stall.
The travel graph below shows Geoff's journey.

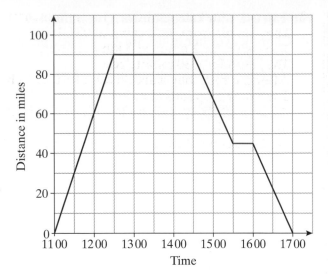

a) How long did Geoff spend at Bradingham Gardens?
b) Work out the speed of the coach, in miles per hour, on the **outward** journey.
c) Did the coach travel at the same speed as this on the **return** journey?

7 The diagram shows a distance–time graph for a train travelling between Farham and Gradel.

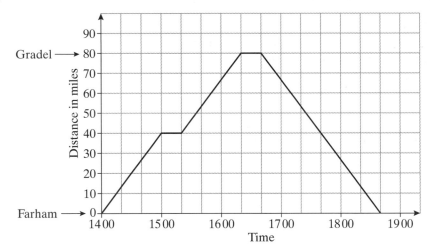

The train leaves Farham at 14 00 for its outward journey to Gradel.
The train leaves Gradel at 16 40 for its return journey to Farham.
a) Work out the speed of the train on its journey from Gradel to Farham.
b) State one difference between the outward journey and the return journey.
c) State one thing that is the same on the outward journey and the return journey.

At 15 00 a second train leaves Gradel.
It travels towards Farham at a constant speed of 40 miles per hour.
d) On a copy of the graph, draw the journey of the second train.
e) At what time do the two trains pass each other?

8 A group of hikers are hiking from their camp to a ruined castle.
They set off from their camp at 08 00
They walk for 2 hours at 3 km per hour.
They stop for a 30 minute rest.
After the rest, they walk on for a further $2\frac{1}{2}$ hours at 4 km per hour.
Then they stop for lunch for 1 hour.
After lunch, they walk on for a further 2 hours at 4 km per hour until they reach the ruined castle.

a) On a copy of the grid, complete the travel graph.

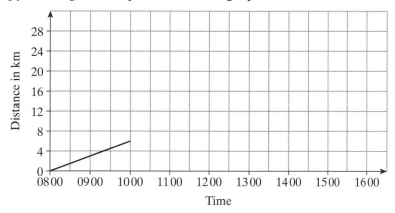

b) How far is the ruined castle from the camp?

9 Zac climbs a steep hill.
He gains height at a rate of 5 metres per minute.
It takes him 1 hour to reach the top.
He then stops for 45 minutes to have lunch.
Then he descends, at 10 metres per minute.

a) How high is the hill?
b) How long does Zac's descent take?
c) Copy and complete the graph below, numbering the scales on both axes.

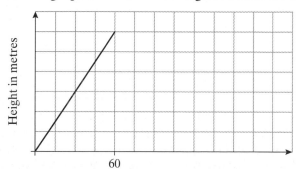

Chapter 13

Angles

1 For each of the following questions
  (i) Work out the angles marked with letters.   **(ii)** Give a reason for your answer.

a)

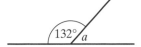

b)

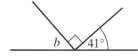

c)

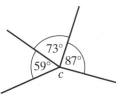

d)

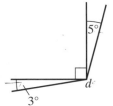

e)

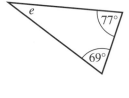

f)

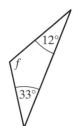

g)

h)

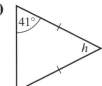

i)

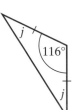

j)

k)

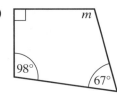

l)

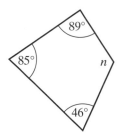

m)

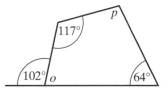

n)

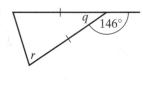

o)

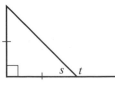

p)

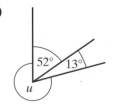

2 Write down what type of angle (e.g. acute, obtuse, reflex) each of the following angles in question **1** are: *c, d, e, f, m, n, s, u.*

EXERCISE 13.2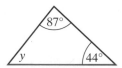

1 Find the value of *y*.

2 Find the value of *m*.

3 a) Write down an expression for the sum of the angles in the triangle. Simplify your expression.
   b) Write down an equation for the sum of the angles in the triangle.
   c) Solve your equation to find the value of *x*
   d) Hence work out the value of each angle.

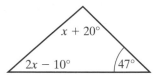

4 a) Write down an expression for the sum of the angles in the triangle. Simplify your expression.
   b) Write down an equation for the sum of the angles in the triangle.
   c) Solve your equation to find the value of *y*
   d) Hence work out the value of each angle.

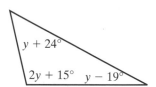

5 A triangle has angles $3x + 16°$, $2x - 26°$ and $90°$
   a) Set up an equation in *x*
   b) Solve your equation, to find the value of *x*
   c) Work out the size of the angles in the triangle.

6 The angles in a triangle are $2y + 10°$, $6y - 32°$ and $4y + 10°$
   a) Set up an equation in *y*
   b) Solve your equation, to find the value of *y*
   c) Work out the size of the angles in the triangle.

7 The diagram shows a quadrilateral.
   Work out the value of *g*

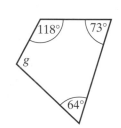

8 The diagram shows a quadrilateral.
   Work out the value of *y*

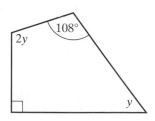

9 Find the size of the angles marked *h*

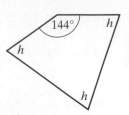

10 a) Form an equation in *y*
   b) Solve your equation to find *y*
   c) Hence find the size of the angles
      in the quadrilateral.

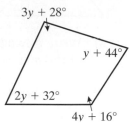

11 a) Form an equation in *k*
   b) Solve your equation to find *k*
   c) Hence find the size of the two angles
      on the straight line.

12 a) Set up an equation in *x*
   b) Solve your equation, to find the value of *x*
   c) Work out the size of all four angles.
   d) Check that your four answers do add up to 360°

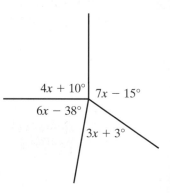

EXERCISE 13.3

Find the size of each of the angles represented by the letters in each question.

1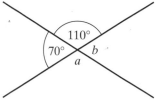

2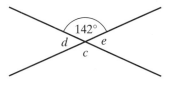

3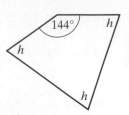

4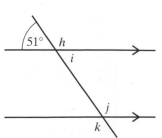

5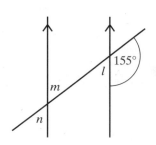

6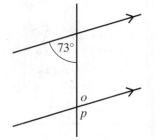

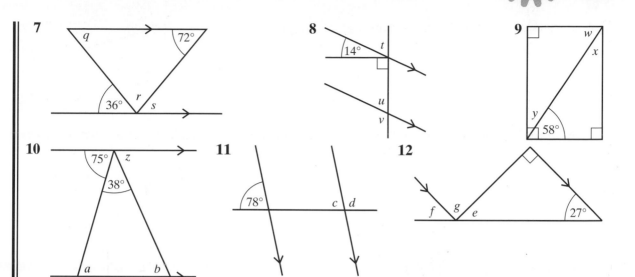

7 q, $72°$, $36°$, r, s

8 $14°$, t, u, v

9 w, x, y, $58°$

10 $75°$, z, $38°$, a, b

11 $78°$, c, d

12 f, g, e, $27°$

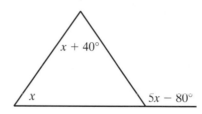

13 a) Form an equation in x.
   b) Solve your equation to find x.

$x + 40°$

x $5x - 80°$

EXERCISE 13.4

1 a) Find the sum of the interior angle at each vertex of:
   (i) a regular hexagon   **(ii)** a regular 10-sided polygon.
   b) Find the size of each interior angle of:
   (i) a regular hexagon   **(ii)** a regular 10-sided polygon.
   c) Do the following tessellate?
   (i) a regular hexagon   **(ii)** a regular 10-sided polygon
   Give a reason for each answer.

2 a) Work out the size of the exterior angle at each vertex of:
   (i) a regular pentagon   **(ii)** a regular 12-sided polygon.
   b) Work out the size of the interior angle at each vertex of:
   (i) a regular pentagon   **(ii)** a regular 12-sided polygon.
   c) Does a:
   (i) regular pentagon   **(ii)** a regular 12-sided polygon tessellate?
   Give a reason for each answer.

3 Seven of the angles in an octagon are $152°$, $123°$, $168°$, $145°$, $110°$, $125°$ and $130°$.
   Find the eighth angle.

4 a) The diagram shows part of a regular polygon.
Work out how many sides the polygon has.

Emma draws this diagram.
She says it shows part of a regular polygon.

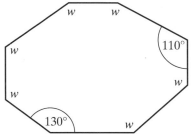

b) Explain how you can tell that Emma must have made a mistake.

5 The diagram shows an irregular hexagon.
Work out the value of h

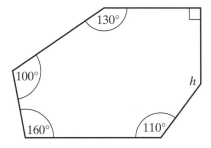

6 The diagram shows an octagon.
All the angles marked w are equal.
Calculate the value of w

7 a) Write down an expression for the sum of the angles in the pentagon.
Simplify your expression.
b) Write down an equation for the sum of the angles in the pentagon.
c) Solve your equation to find the value of x
d) Hence work out the value of each angle.

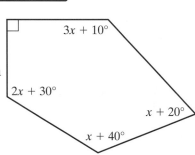

8 Follow these instructions to make an accurate drawing of a regular pentagon.

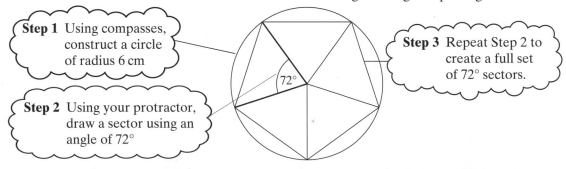

Step 1 Using compasses, construct a circle of radius 6 cm

Step 2 Using your protractor, draw a sector using an angle of 72°

Step 3 Repeat Step 2 to create a full set of 72° sectors.

Complete the construction by joining the 5 points around the circumference of the circle.

9 Adapt the instructions from question **8** to make:
a) a regular hexagon
b) a regular 10-sided polygon
c) a 10-pointed star.

1 Measure the bearings of each of the following six towns from Gingbridge.

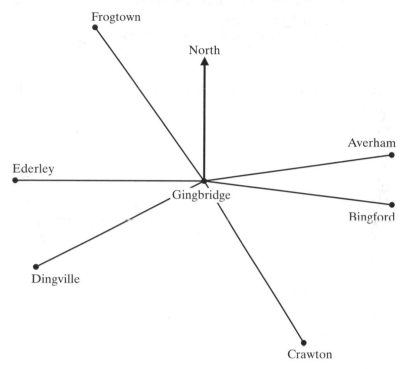

2 Write down the bearing on which a plane flies when it is flying:
   a) due south   **b)** due east   **c)** north-east   **d)** south-west.

3 Huntville is 9 km away from Ivaton on a bearing of 065°
   Make a scale drawing of the two towns. Use a scale of 1 cm to 1 km

4 A ship is 10.8 km away from a lighthouse on a bearing of 250°
   a) Make a scale drawing of the ship and the lighthouse. Use a scale of 1 cm to 1 km

   A ferry is 15 km away from the ship on a bearing of 127°
   b) Add the position of the ferry to your scale drawing.
   c) Measure the bearing of the ferry from the lighthouse.

5 Look at this sketch. It is not drawn to scale.
   a) Write down the bearing of B from A.
   b) Work out the bearing of A from B.

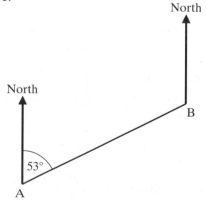

6 Look at this sketch. It is not drawn to scale.
  a) Work out the bearing of R from P.
  b) Work out the bearing of P from R.

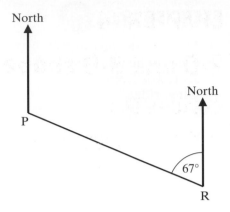

7 Abbeyton is north-west of Bonville.
  a) What is the bearing of Abbeyton from Bonville?
  b) What is the bearing of Bonville from Abbeyton?

height

8 A helicopter and a plane are flying at the same altitude.
The bearing of the helicopter from the plane is 195°
What is the bearing of the plane from the helicopter?

9 Cogtown is on a bearing of 320° from Damston.
What is the bearing of Damston from Cogtown?

10 This sketch shows the position of three boats.
It is not a scale drawing.

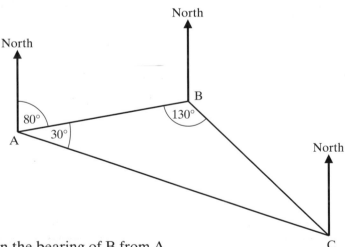

  a) Write down the bearing of B from A.
  b) Work out the bearing of:
      (i) A from B **(ii)** C from A **(iii)** B from C **(iv)** A from C **(v)** C from B.

CHAPTER 14

2-D and 3-D shapes

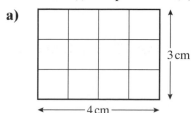

EXERCISE 14.1

1 Name the types of triangles and quadrilaterals labelled A to M in the diagram.

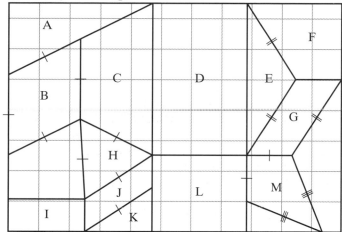

2 Two different types of quadrilaterals have all four sides the same length.
What types of quadrilaterals are these?

3 Three different quadrilaterals always have exactly two pairs of sides the same length.
What types of quadrilaterals are these?

4 A quadrilateral has all four sides the same length.

Is Mel correct?
Explain why?

All four angles must be the same size.

Mel

EXERCISE 14.2

1 Work out: **(i)** the perimeter, **(ii)** the area of each of these rectangles.

a)

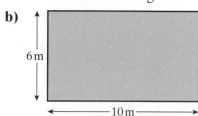

3 cm

4 cm

b)

6 m

10 m

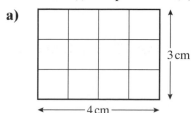

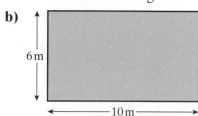

c)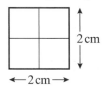
4 mm
9 mm

d)
28 cm
4 cm

You can use your calculator for Questions 2–10.

2 Work out: **(i)** the perimeter, **(ii)** the area of each of these squares.

a)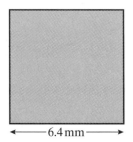
2 cm
2 cm

b)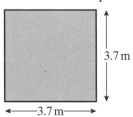
3.7 m
3.7 m

c)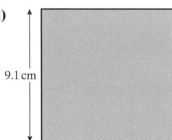
6.4 mm

d)
9.1 cm

3 Work out: **(i)** the perimeter, **(ii)** the area of each of these triangles.

a)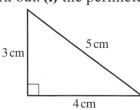
3 cm
5 cm
4 cm

b)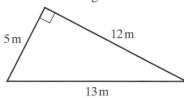
5 m
12 m
13 m

c)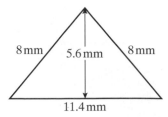
8 mm
5.6 mm
8 mm
11.4 mm

d)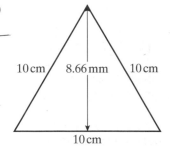
10 cm
8.66 mm
10 cm
10 cm

4 Work out the area of these triangles.

a)
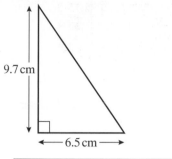
9.7 cm
6.5 cm

b)
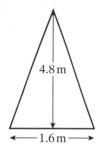
4.8 m
1.6 m

c)
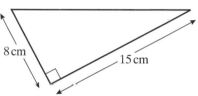
8 cm
15 cm

d)
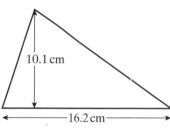
10.1 cm
16.2 cm

e)
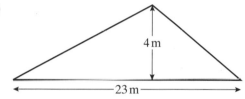
4 m
23 m

f)
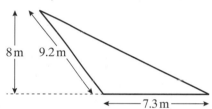
8 m   9.2 m
7.3 m

5 Find: **a)** the area, **b)** the perimeter of this triangle.

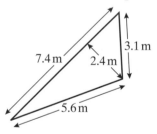
7.4 m   2.4 m   3.1 m
5.6 m

6 A rectangle has an area of 18 cm^2
Its sides are a whole number of centimetres.
a) On squared paper, draw three different rectangles
with an area of 18 cm^2

18 cm^2

The perimeter of the rectangle is 22 cm
b) What are the dimensions of the rectangle?

7 A square has a perimeter of 40 cm
What is the area of the square?

The length
and width.

8 A square has an area of 81 cm^2
What is the perimeter of the square?

9 This classroom has an area of $60\,m^2$
a) What is the length of the classroom?
b) What is the perimeter of the classroom?

5 m

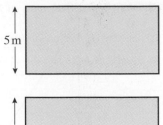

10 A farmer puts 180 m of fencing around a rectangular field.
a) What is the length of the field?
b) What is the area of the field?

35 m

EXERCISE 14.3

1 The diagram shows a rectangle.
The perimeter of the rectangle is 60 cm
a) Write down an expression for the perimeter of the rectangle.
Simplify your expression.
b) Write down an equation in x for the perimeter of the rectangle.
c) Solve your equation to find the value of x.
d) Find the length and width of the rectangle.

$x + 9$

x

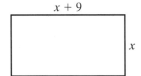

2 The diagram shows an isosceles triangle. AB = AC. All the lengths are in cm.

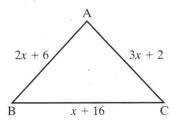

A

$2x + 6$ $3x + 2$

B $x + 16$ C

a) Use the information that AB = AC to set up an equation in x
b) Solve your equation to find the value of x
c) Hence work out the perimeter of the triangle.

3 The diagram shows a triangle. All the lengths are in cm.

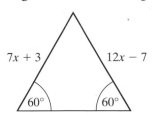

$7x + 3$ $12x - 7$

60°   60°

a) What type of triangle is this?
b) Set up, and solve, an equation in x
c) Hence work out the perimeter of the triangle.

EXERCISE 14.4

Do not use your calculator for Questions 1 and 2.

1 Find the area of each of these parallelograms:

a)

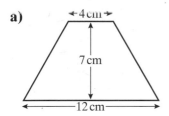

5 cm
12 cm

b)

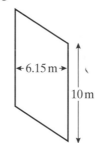

6.15 m
10 m

2 Find the area of each of these trapeziums:

a)

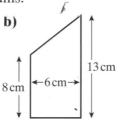

4 cm
7 cm
12 cm

b)

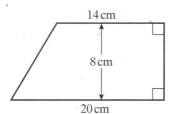

13 cm
8 cm
6 cm

You can use your calculator for Questions 3 and 4.

3 Find the area of each shape.

a)

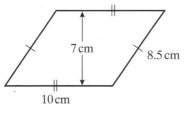

7 cm
8.5 cm
10 cm

b)

14 cm
8 cm
20 cm

c)

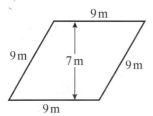

9 m
9 m
7 m
9 m
9 m

d)

5.6 cm
4 cm
3.8 cm
4 cm
8.2 cm

4 The diagram shows a parallelogram.
All lengths are in metres.

a) Set up, and solve, an equation in x
b) Set up, and solve, an equation in y
c) Hence work out the lengths of the
sides of the parallelogram.

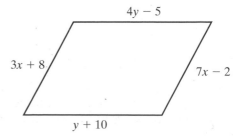

$4y - 5$
$3x + 8$
$7x - 2$
$y + 10$

14 2-D and 3-D shapes   77

EXERCISE 14.5

1 Find the shaded area of each shape.

a)

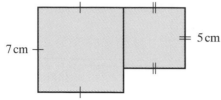

7 cm · 5 cm

b)

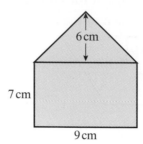

6 cm

7 cm

9 cm

c)

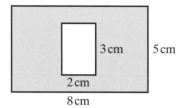

3 cm · 5 cm · 2 cm · 8 cm

d)

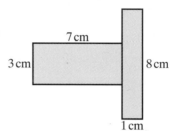

7 cm · 3 cm · 8 cm · 1 cm

e)

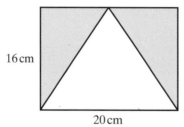

16 cm · 20 cm

f)

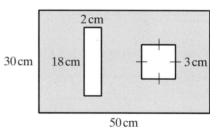

2 cm · 30 cm · 18 cm · 3 cm · 50 cm

You can use your calculator for Questions 2 and 3.

2 Calculate: **(i)** the perimeter, **(ii)** the area of each shape. State the units in each case.

a)

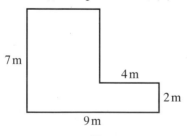

7 m · 4 m · 2 m · 9 m

b)

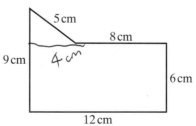

5 cm · 8 cm · 9 cm · 4 cm · 6 cm · 12 cm

c)

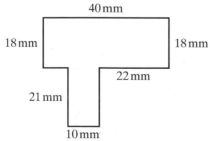

40 mm · 18 mm · 18 mm · 22 mm · 21 mm · 10 mm

d)

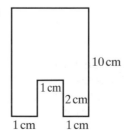

10 cm · 1 cm · 2 cm · 1 cm · 1 cm

e)

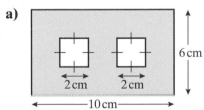

7 m
5 m
3 m
3 m
3 m

f)

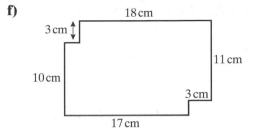

18 cm
3 cm
11 cm
10 cm
3 cm
17 cm

3 Find the shaded area of each shape.

a)

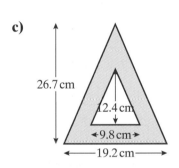

6 cm
2 cm
2 cm
10 cm

b)

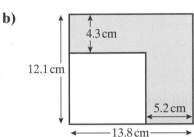

4.3 cm
12.1 cm
5.2 cm
13.8 cm

c)

26.7 cm
12.4 cm
9.8 cm
19.2 cm

d)

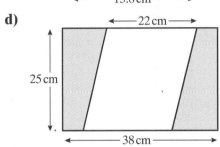

22 cm
25 cm
38 cm

EXERCISE 14.6 pt

Do not use your calculator for Questions 1–3.

1 Find:
   (i) the volume,
   (ii) the surface area of these shapes.

a)   **b)**   **c)**

represents
1 cm^3

2 The diagram shows a cube of side 3 m
Calculate its   **a)** surface area,   **b)** volume.
State the units in your answers.

3 m
3 m
3 m

3 The diagram shows a cuboid, with
dimensions 5 cm, 8 cm and 10 cm
Calculate its   **a)** surface area,   **b)** volume.
State the units in your answers.

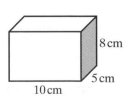

8 cm
5 cm
10 cm

You can use your calculator for Questions 4–11.

4 The diagram shows a prism.
Its cross section is a right-angled triangle of sides
6 cm, 8 cm, 10 cm
The prism has a length of 12 cm
   a) Calculate the area of the cross section, shaded
      in the diagram.
   b) Work out the volume of the prism.
   c) Calculate the surface area of the prism.

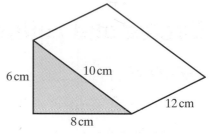

5 The cross section of a steel girder is in the shape of a letter T.
The cross section is shown in the diagram.
   a) Work out the area of the T-shaped cross section.

   The girder is 50 cm long.
   b) Work out the volume of the girder.

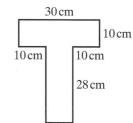

6 A cube measures 20 cm along each side.
   a) Work out the volume of the cube.
   b) Work out the surface area of the cube.

7 A cuboid measures 37 cm by 40 cm by 5 cm
   a) Work out the volume of the cuboid.
   b) Work out the surface area of the cuboid.

8 The diagram shows a water tank.
It is in the shape of a cuboid.
It has no lid.
   a) Work out the volume of the tank,
      correct to 3 s.f.
   b) Work out the total surface area of the
      inside of the tank.

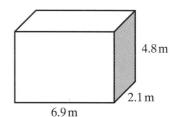

9 The diagram shows a sketch
of a swimming pool.

The pool is 1 m deep at the shallow end,
and 3 m deep at the deep end.
The pool is 25 m long and 8 m wide.
   a) Work out the volume of the pool.

   1 cubic metre = 1000 litres
   b) Work out the number of litres of water in the pool when it is full.

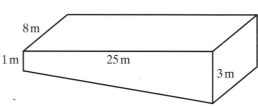

10 A cube has a volume of 1000 cm³
   a) Work out the dimensions of the cube.
   b) Calculate the surface area of the cube.

11 A cuboid has a volume of 40 cm³
Its dimensions are all integers bigger than 1
No two dimensions are the same.
   a) Work out the dimensions of the cuboid.
   b) Calculate the surface area of the cuboid.

14 2-D and 3-D shapes

CHAPTER 15

Circles and cylinders

EXERCISE 15.1

Give the answers to each of these problems correct to 3 significant figures.

1 Find the circumference of these circles:

a) 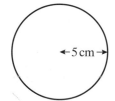 ←5 cm→

b) ←12 mm→

c) 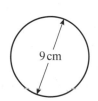 9 cm

d) 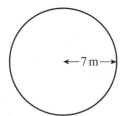 ←7 m→

2 Find the area of these circles:

a) 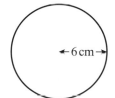 ←6 cm→

b) ←4 m→

c) ←1 cm→

d) 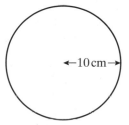 ←10 cm→

3 A circle has radius 11 mm
Find its circumference.

4 A circle has diameter 30 cm
Find its circumference.

5 A circle has radius 15 cm
Find its area.

6 A circle has diameter 42.8 cm
Find its area.

7 A circle has diameter 13 cm
Find its area.

8 A circle has radius 0.43 m
Find its area.

9 A circle has diameter 370 cm
Find its circumference.

10 A circle has radius 8.05 m
Find its circumference.

EXERCISE 15.2 pt

Give the answers to each of these problems correct to 3 significant figures.

1 A circle has a diameter of 24 mm. Calculate its circumference.

2 A face of a coin is a circle of diameter of 32.4 mm. Calculate its area.

3 Anna decides to run around a circular race track.
   The radius of the track is 30 metres.
   a) Work out the length of one lap of the track.

   Anna wants to run at least 2000 metres.
   She wants to run a whole number of laps.
   b) Work out the minimum number of laps that Anna must run.

4 The diagram shows a rectangle inside a circle.
   The circle has a radius of 50 cm
   The rectangle measures 80 cm by 60 cm
   The corners of the rectangle are on the circumference of the circle.
   a) Work out the area of the rectangle.
   b) Work out the area of the circle.
   c) Work out the shaded area.
   Give your answer correct to the nearest square centimetre.

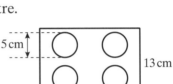

5 The diagram shows a rectangular piece of card with
   four circles cut out of it.
   The circles are each of diameter 5 cm
   The rectangle measures 24 cm by 13 cm
   a) Calculate the area of one of the circular holes.
   b) Work out the area of the card.

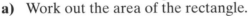

6 A circular table stands on a circular base.
   The diameter of the table top is 94 cm
   The radius of the base is 33 cm
   a) Calculate the area of the table top.
   b) Calculate the circumference of the base.

7 A bicycle wheel has a diameter of 70 cm
   a) How far, in metres, does the wheel go when it does one complete revolution?
   b) How far, in metres, does the wheel go when it does 20 complete revolutions?
   c) How many revolutions does the wheel do when it travels it 80 m?

8 The diagram shows a circular garden with a circular pond in the middle.
   The whole garden, apart from the pond, is lawn.
   The garden has a radius of 5 m
   The pond has a diameter of 2 m
   a) Calculate the area of the pond.
   b) Calculate the area of the garden.
   c) Hence find the area of the lawn.

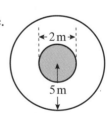

   The gardener wants to put some fertiliser on the lawn.
   One bag of fertiliser is sufficient for 0.8 m²
   d) Work out how many bags of fertiliser the gardener needs to buy.

EXERCISE 15.3

Give the answers to each of these problems correct to 3 significant figures.

1 Calculate **(i)** the perimeter, **(ii)** the area of each sector.

a)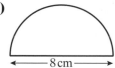
8 cm

b)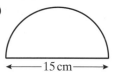
15 cm

c)
5 cm

d)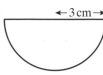
←3 cm→

e)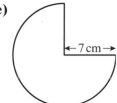
←7 cm→

f)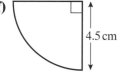
4.5 cm

g)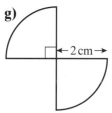
←2 cm→

h)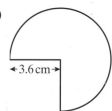
←3.6 cm→

2 A pizza of diameter 16 inches is to be shared between two people.
The pizza is cut into two equal sectors.
Work out the area of pizza each person receives.

3 The diagram shows a running track at a school.
It is made up of two straight sections,
and two semicircular ends.
The dimensions are marked on the diagram.

Andy runs around the outside boundary of the
track, marked with a solid line.
Colin runs around the inside boundary, marked
with a dotted line.
They each run one lap of the track.
a) Work out how far Colin runs.
b) Work out how much further than Colin, Andy runs.

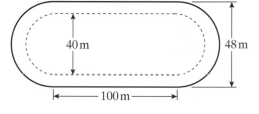
40 m   48 m
100 m

4 Work out the total area of the shaded segments
in the diagram on the right.

6 cm
←6 cm→←6 cm→

EXERCISE 15.4

Give the answers to each of these problems correct to 3 significant figures.

1 A circle has circumference 46.7 cm
Find its diameter.

2 A circle has circumference 38.2 cm
Find its diameter.

3 A circle has circumference 55 m
Find its diameter.

4 A circle has circumference 28 mm
Find its radius.

5 A circle has circumference 39 cm
Find its radius.

6 A circle has circumference 18 m
Find its radius.

7 A circle has circumference 60 cm
Find its radius.

8 A circle has circumference 3 m
Find its radius.

9 The diagram shows a running track.
The ends are semicircles of radius y metres.
The straights are of length 45 metres each.
The total distance around the track
is 120 metres.

Calculate the value of y

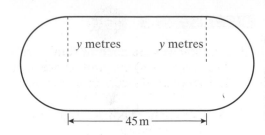

45 m

EXERCISE 15.5

Give the answers to each of these problems correct to 3 significant figures.

1 A cylinder has radius 15 cm and height 23 cm
Find its volume.

2 A cylinder has radius 4 cm and height 3 cm
Find its curved surface area.

3 A cylinder has diameter 36 cm and height 10 cm
a) Find its volume.
b) Find its curved surface area.

4 A cylinder has radius 7 cm and height 18 cm
Find its volume.

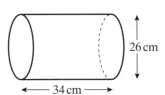

26 cm

5 The diagram on the right shows a hollow cylinder.
Work out the curved surface area of the cylinder.

34 cm

6 A hollow cylindrical pipeline has an internal diameter of 16 cm
The pipeline is 250 metres in length.
a) Work out the volume of the pipeline. Give your answer in cm^3
b) $1000 \, cm^3 = 1$ litre. Express the volume of the pipeline in litres.

7 A tube of crisps is in the shape of a cylinder.
It has radius 7 cm and height 30 cm
Work out the volume of the cylinder.

8 A sweet packet is in the shape of a hollow cardboard cylinder.
The inside diameter of the cylinder is 4.5 cm and it has a height of 12 cm
a) Work out the volume of the cylinder.

The sweets have a volume of 2.3 cm^3 each.
b) Show that the packet cannot contain as many as 90 sweets.

9 Rosie wants to buy a can of soup.
She wants to buy the can of soup that
has the greatest volume.

Which of these two cans should she buy?
Show all your working.

A

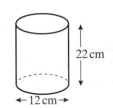

22 cm

12 cm

B

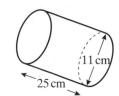

11 cm

25 cm

EXERCISE 15.6

1 A circle has diameter 30 cm
Work out its **a)** circumference, **b)** area.
Leave your answers in terms of π.

2 A circle has radius 18 cm
Work out its **a)** circumference, **b)** area.
Leave your answers in terms of π.

3 A cylinder has radius 10 cm and
height 7 cm Work out its:
a) curved surface area, **b)** volume.
Leave your answers in terms of π.

4 A circle has circumference 39π cm
Work out **a)** the exact radius,
b) the exact area of the circle.
Leave your answers in terms of π.

5 A circle has a circumference of 5π cm
a) Work out the exact radius of the circle.
b) Work out the exact area of the circle.
Leave your answers in terms of π.

6 A cylinder has volume 580π cm^3
It has radius 20 cm
Work out its height.

7 A cylinder has volume 360π cm^3
It has radius 10 cm
Work out its height.

8 The diagram shows a quadrant of a circle. The radius is 8 cm
a) Work out the area of the quadrant, in terms of π
b) Find an exact expression for the perimeter of this quadrant.

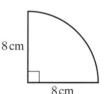

9 The diagram shows an unusual carpet.
It is in the shape of a square, with semicircles on each of
the four sides. The square is of side 20 m
a) Work out the area of one of the semicircles.
Leave your answer in terms of π
b) Find an exact expression for the total area of the carpet.

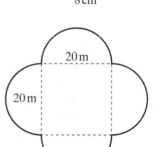

10 The diagram shows two cylinders.
Cylinder A has diameter 5 cm and height 12 cm
Cylinder B has diameter 12 cm and height 5 cm

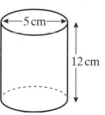

Cylinder A

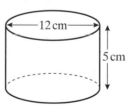

Cylinder B

a) Show that both cylinders have exactly the same curved surface area.
b) Work out the volume of each cylinder, leaving your answers in terms of π
c) Which cylinder has the larger volume?

CHAPTER 16

Pythagoras' theorem

EXERCISE 16.1

1 Look at these triangles, and use Pythagoras' theorem to decide whether they are right angled or not. The diagrams are not drawn to scale.

a)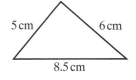
5 cm   6 cm
8.5 cm

b)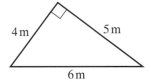
4 m   5 m
6 m

c)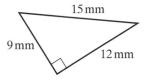
15 mm
9 mm   12 mm

For each of the triangles described below, use Pythagoras' theorem to decide whether it is right angled or not. If so, name the angle at which the right angle is located.

2 AB = 9 cm   BC = 7 cm   CA = 5.4 cm

3 AB = 2.5 cm   BC = 1.5 cm   CA = 2 cm

4 AB = 8 mm   BC = 8 mm   CA = 11 mm

5 PQ = 2.6 cm   QR = 3.5 cm   RP = 5 cm

6 PQ = 2.5 m   QR = 6 m   RP = 6.5 m

7 PQ = 7.2 cm   QR = 8.3 cm   RS = 12 cm

8 AB = 15 km   BC = 20 km   CA = 25 km

EXERCISE 16.2

1 Find the length of the hypotenuse represented by the letters *a* to *f* below. Give your answers to 3 significant figures.

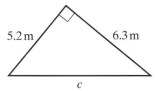
3 cm   *a*   7 cm

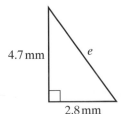
b   2 cm   8 cm

5.2 m   6.3 m   *c*

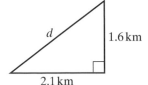
d   1.6 km   2.1 km

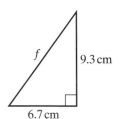
4.7 mm   *e*   2.8 mm

f   9.3 cm   6.7 cm

2 Find the length of the diagonal of each rectangle.

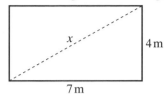

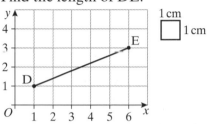

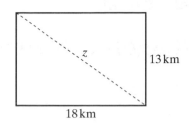

3 a) Find the length of **(i)** AB, **(ii)** BC.
  b) Hence find the length of AC.

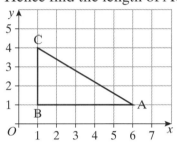

4 Find the length of DE.

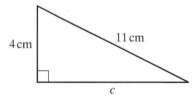

EXERCISE 16.3 pt

1 Find the length of the sides marked a to f below.
   Give your answers to 3 significant figures.

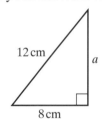

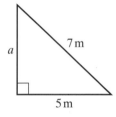

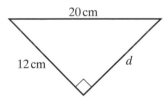

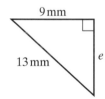

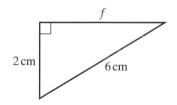

2 A rectangle has length 6 m and width x cm
   Its diagonal is of length 10 m
   Find the value of x

3 A ship sails due North for 8 km, then turns and sails due East for y km
   It ends up 15 km in a direct straight line from its starting point.
   Find the value of y

EXERCISE 16.4

1 Find the length of the side marked by the letters *a* to *f* below.
Give your answers to 3 significant figures where appropriate.

a)
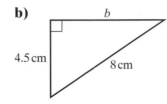
a
3.6 cm
5.2 cm

b)
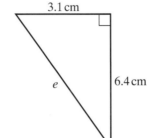
b
4.5 cm
8 cm

c)
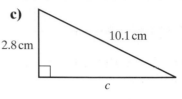
2.8 cm
10.1 cm
c

d)

10 cm   10 cm
d

e)
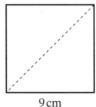
3.1 cm
e   6.4 cm

f)
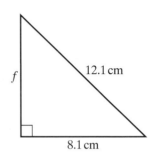
f   12.1 cm
8.1 cm

2 Find the length of the diagonals in these squares.

a)
7 cm

b)
9 cm

3 a) Work out the height of this isosceles triangle.

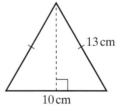
13 cm
10 cm

b) Find the area of the triangle.

4 The diagram shows a rectangular field with a footpath running diagonally across it.
a) Work out the length of the footpath.

Ahmed walks from A to B and then from B to C.
Anthony walks from A to C along the footpath.
b) How much farther does Ahmed walk?

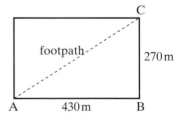
C
footpath
270 m
A   430 m   B

5 A man places a ladder against a wall.
The foot of the ladder is 3.4 m from the wall.
The top of the ladder is 2.7 m from the ground.
Work out the length of the ladder.

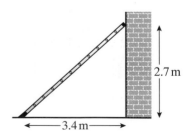
2.7 m
3.4 m

CHAPTER 17

Transformations

EXERCISE 17.1

1 (i) Draw all the lines of symmetry on a copy of the following shapes.
   (ii) Write down how many lines of symmetry each shape has.

a)

Rectangle

b)

Isosceles triangle

c)

Rhombus

d)

Parallelogram

e)

Trapezium

f)

Regular octagon

2 (i) Draw all the lines of symmetry on a copy of the following shapes.
   (ii) Write down how many lines of symmetry each shape has.

a)
b)
c)
d)

3 Shade in squares on a copy of the following grids so that the resulting pattern is symmetrical about the lines shown.

a)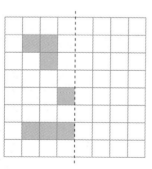

b)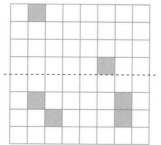

c)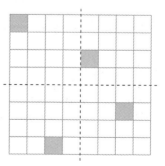

(continued)

d) 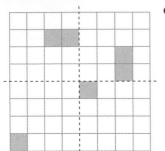   **e)** 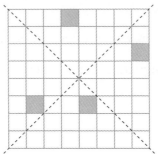   **f)**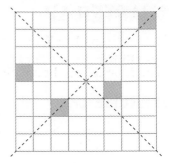

4 Sketch in a plane of symmetry on a copy of each of the following shapes:

a)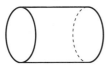

b)

c)

d)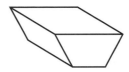

5 Which of these shapes are **congruent**?

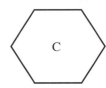
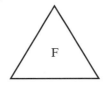
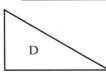

A

B

C

D

E   F   G

H   I   J   N

K   L   M

EXERCISE 17.2

In Questions **1** to **7**, on a copy draw the reflection of the given shape in the mirror line shown.
Label the mirror line with its equation in each case.

1

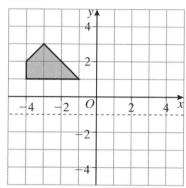

2

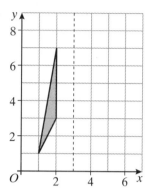

3

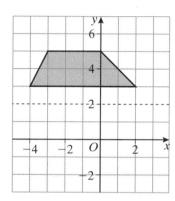

4

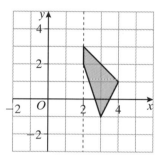

5

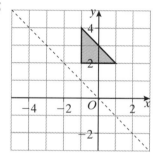

6

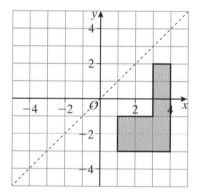

7

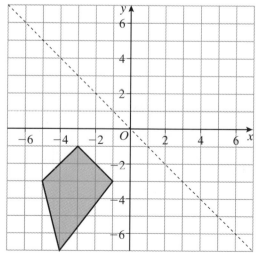

8 The diagram shows a triangle P and its mirror image Q.

   a) On a copy of the diagram, draw the mirror line that has been used for the reflection.
   b) Write down the equation of the mirror line.

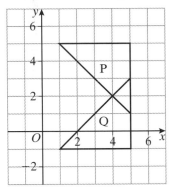

9 The diagram shows a triangle S and its mirror image T

   a) On a copy of the diagram, draw the mirror line that has been used for the reflection.
   b) Write down the equation of the mirror line.

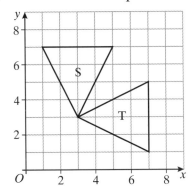

10 The diagram shows a letter T shape, labelled X.
   The shape is to be reflected in a mirror line.
   Part of the reflection has been drawn on the diagram.

   a) On a copy of the diagram, complete the drawing to show the image.

       Label it Y.
   b) Mark the mirror line.
   c) Give its equation.

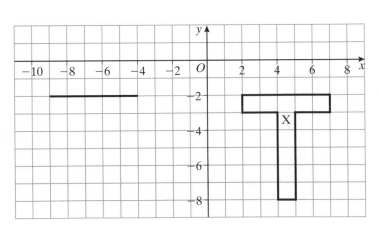

11 The diagram shows ten triangles A, B, C, D, E, F, G, H, I and J.
The ten triangles are all congruent to each other.

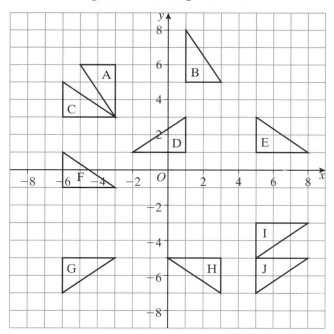

a) Explain the meaning of the word *congruent*.

b) Triangle E can be reflected to triangle I.
State the mirror line that achieves this.

c) Triangle H is reflected to another triangle using the mirror line $x = -1\frac{1}{2}$
Which one?

d) Triangle E can be reflected to triangle D using a mirror line.
Give the equation of this line.

e) Triangle B can be reflected to triangle E using a mirror line.
Give the equation of this line.

f) Triangle A can be reflected to triangle C using a mirror line.
Give the equation of this line.

g) Triangle G is reflected to another triangle using the mirror line $y = -3$.
Which one?

12 A triangle P is reflected in the line
$x = 3$, to form an image, triangle Q.
Then triangle Q is reflected in the
same mirror line, to form an image,
triangle P.
What can you say about triangle P
and triangle Q?
Use a copy of the grid to help you.

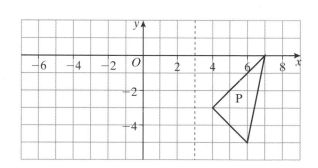

In Questions **1–4**, describe: **a)** the translation that takes A onto B
b) the translation that takes B onto A.

1
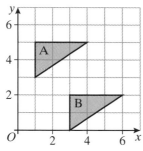

2
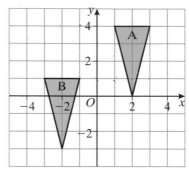

3
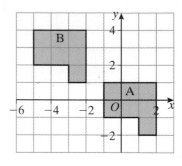

4
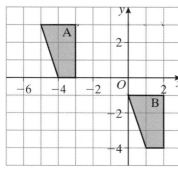

5
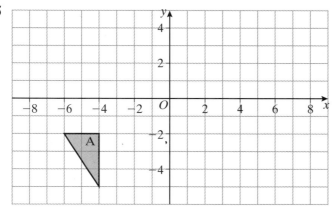

a) On a copy of the diagram, translate triangle A by the vector $\begin{pmatrix} 4 \\ 2 \end{pmatrix}$. Label the image B.

b) Translate triangle B by the vector $\begin{pmatrix} 5 \\ -1 \end{pmatrix}$. Label the image C.

c) Translate triangle C by the vector $\begin{pmatrix} -7 \\ 5 \end{pmatrix}$. Label the image D.

d) Describe the single translation that would take triangle D to triangle A.

6

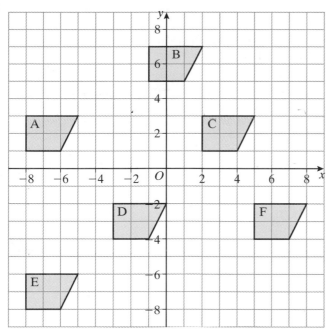

Describe the translation that takes the trapezium from:
a) A onto B
b) D onto F
c) D onto E
d) E onto A
e) E onto C
f) C onto B
g) A onto F
h) B onto D

EXERCISE 17.4

1 Write down the order of rotational symmetry of each of these shapes:

a)
b)
c)
d)

e)
f)
g)
h)

2 On a copy of this diagram, complete this shape so that is has two lines of symmetry and rotational symmetry of order 2

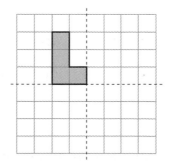

3 On two copies of this diagram, add
one square to this shape so it has
rotational symmetry of order 2
There are two possible solutions.
Draw each solution on a separate grid.

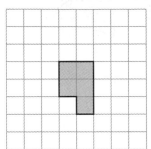

4 On a copy of this diagram, add three
squares to this shape so that it has
rotational symmetry of order 4

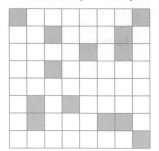

EXERCISE 17.5

Use copies of the coordinate grids with *x* and *y* axes numbers from −8 to 8.

1 Rotate the trapezium shape 90°
clockwise, about *O*.

2 Rotate the L-shape by 180°, about *O*.

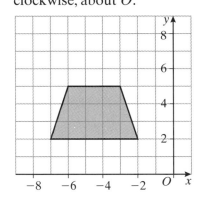

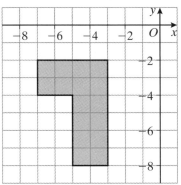

3 a) Rotate triangle T1 90° anticlockwise about *O*.
Label the result T2.
 b) Rotate T2 180° about *O*.
Label the result T3.
 c) Describe the single rotation that takes T1
directly to T3.

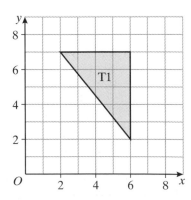

4 **a)** Rotate triangle A 90° anticlockwise about $(0, 0)$.
   Label the result B.
   b) Rotate triangle B 180° about $(0, 0)$.
   Label the result C.
   c) Describe the single rotation that takes
   triangle C to triangle A.

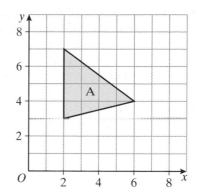

5

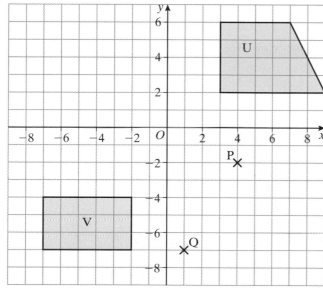

a) Rotate shape U 90° anticlockwise about point P $(4, -2)$
b) Rotate shape V 90° clockwise about point Q $(-7, 1)$

6

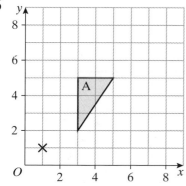

a) Rotate triangle A 90° clockwise about $(1, 1)$. Label it B.
b) Now rotate both triangle A and triangle B each by 180° about $(1, 1)$

7 The diagram shows an object A and its image B after a rotation.

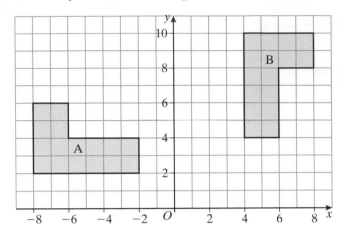

a) Write down the size and direction of the angle of rotation.
b) Write down the coordinates of the centre of rotation.

EXERCISE 17.6

Use copies of the coordinate grids with *x* and *y* axes numbered from −8 to 8.

1 a) Reflect triangle S in the line $x = -1$
Label the new triangle T.
b) Reflect triangle T in the *x* axis.
Label the new triangle U.

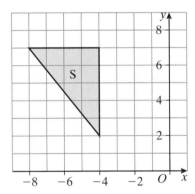

2 The diagram shows a triangle, T.

a) Translate triangle T by $\begin{pmatrix} 0 \\ 4 \end{pmatrix}$

Label its image U.

b) Rotate triangle U by 90° clockwise about *O*.
Label the resulting triangle V.

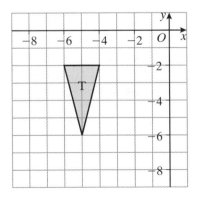

17 Transformations

3 The diagram shows a triangle S.

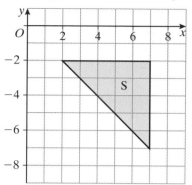

a) Reflect triangle S in the y axis.
   Label this image T.
b) Reflect triangle T in the line $y = -1$
   Label this image U.

4 The diagram shows a rectangle.
   The rectangle is labelled F1.

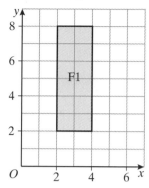

a) Reflect the rectangle F1 in the
   y axis.
   Label the result F2.
b) Reflect F2 in the line $y = x$.
   Label the result F3.

5 The diagram shows a quadrilateral E.

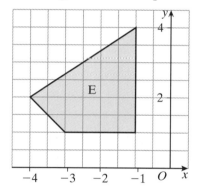

a) Rotate quadrilateral E through $90°$
   clockwise about O.
   Label the result F.
b) Rotate quadrilateral F through $90°$
   anticlockwise about $(1, -1)$
   Label the result G.
c) Describe the translation that would
   take shape G to shape E.

6 The diagram shows a triangle P.

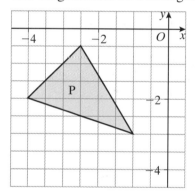

a) Rotate triangle P by $90°$
   anticlockwise about $(0, 1)$.
   Label the result Q.
b) Reflect triangle P in the x axis.
   Label the image R.
c) Describe the translation that
   takes triangle R to triangle Q.

EXERCISE 17.7

Copy the following shapes in this Exercise onto squared paper.

1 Enlarge each shape by scale factor 2

   a)   **b)**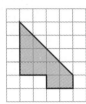

2 Enlarge the following shapes by scale factor 3

   a)   **b)**

3 Enlarge the following shapes by scale factor $\frac{1}{2}$

   a) 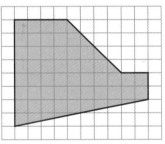   **b)**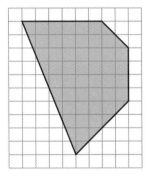

4 Enlarge the following shape by scale factor:

   a) $\frac{1}{2}$ **b)** $1\frac{1}{2}$ **c)** 2   **d)** $2\frac{1}{2}$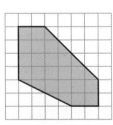

5 A shape has a perimeter of 16 cm
Write down the perimeter of the shape when it is enlarged by a scale factor of:
   a) 2   **b)** 4   **c)** $1\frac{1}{2}$ **d)** $\frac{1}{2}$

6 Write down which of these rectangles are:
  a) similar to
   (i) A   **(ii)** B   **(iii)** C
  b) congruent (there are three pairs).

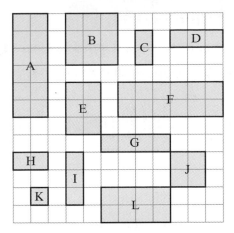

7 Which of these triangles are similar to triangle A?

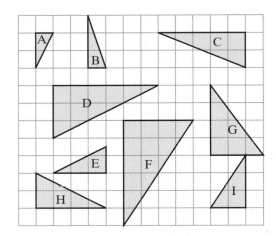

8 Are any two equilateral triangles similar?
  Give a reason for your answer.
  Use this diagram to help you.

9 Are any two rhombuses similar?
  Give a reason for your answer.
  Use this diagram to help you.

10 Are any two regular pentagons similar?
  Give a reason for your answer.
  Use this diagram to help you.

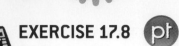

EXERCISE 17.8 pt

For Questions 1, 2 and 3 in this Exercise, draw a copy of the grid.

1 The diagram shows a shape A.

a) Enlarge shape A by scale factor 2, centre P.
Label the new shape B.

b) Enlarge shape A by scale factor 3, centre P.
Label the new shape C.

c) State whether shapes B and C are:
(i) congruent
(ii) similar.

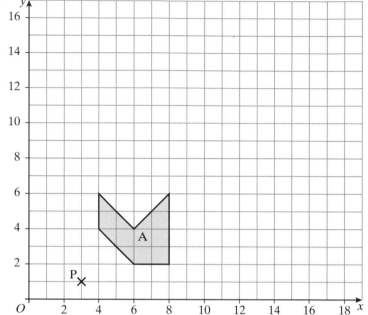

2 The diagram shows a trapezium.

Enlarge the shape by scale factor $2\frac{1}{2}$, using centre $(6, -8)$

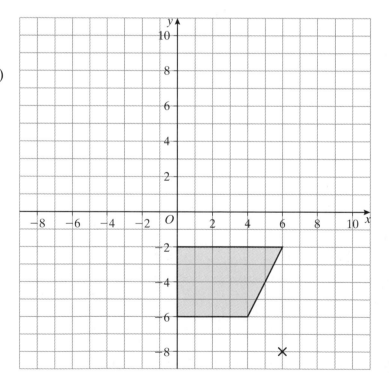

3 The diagram shows a shape, P, and two centres A and B marked with crosses.

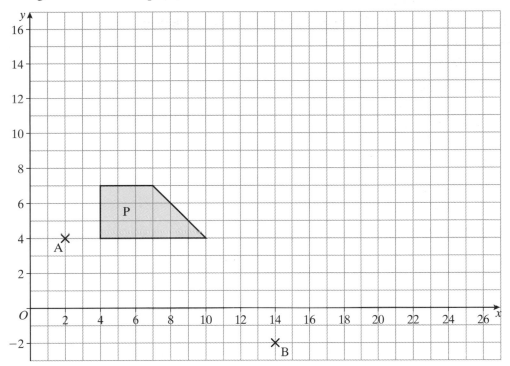

a) Enlarge shape P, with scale factor 3, centre A (2, 4)
Label the result Q.

b) Enlarge shape Q, with scale factor $\frac{1}{3}$, centre B (14, −2)
Label the result R.

c) State whether shapes P and Q are: **(i)** congruent   **(ii)** similar.

d) Are shapes P and R congruent?

4 The diagram shows an object, A, and the result B after an enlargement.

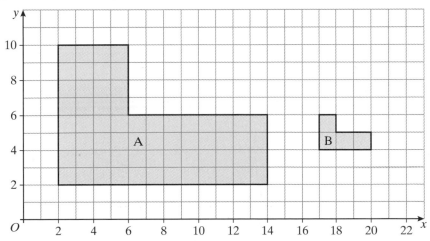

State the scale factor for the enlargement.

CHAPTER 18

Constructions and loci

EXERCISE 18.1

1 Use ruler, compasses and protractors to construct these triangles.
The diagrams are *not* accurately drawn.

a)

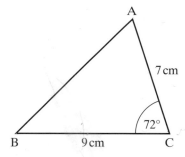

b)

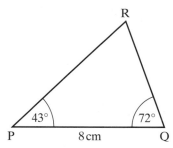

c)

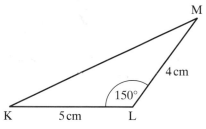

d)

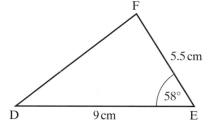

e)

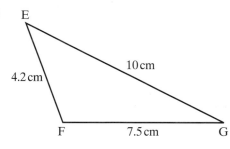

f)

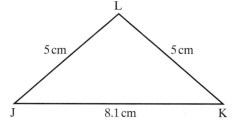

2 The sketch shows a triangle with
AB = 73 mm, BC = 56 mm, AC = 87 mm
Using ruler and compasses, make an
accurate diagram of the triangle.

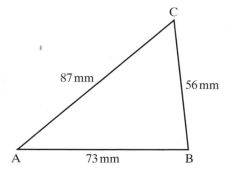

3 The sketch shows a triangle with AB = 84 mm,
BC = 52 mm and angle ACB = 90°
Using ruler and compasses, make an accurate
diagram of the triangle.

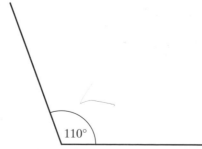

C

52 mm

A   84 mm   B

4 Draw triangle PQR with PQ = 8.1 cm, QR = 8.8 cm, angle RPQ = 90°
Use ruler and compasses to construct this triangle.
Show that there are two different solutions based in the given information.

5 Using compasses, try to make an accurate drawing of triangle EFG with sides
EF = 14 cm, EG = 6 cm, FG = 7 cm
What difficulty do you encounter? Find a reason for this.

EXERCISE 18.2

1 Use ruler and compasses to construct the perpendicular bisectors of these line segments.

a) A ————— 7 cm ————— B

b) A ————— 12 cm ————— B

c) A ——— 4 cm ——— B

d) A ————— 9 cm ————— B

e) A ————— 8.5 cm ————— B

2 Use ruler and compasses to construct the bisectors of these angles.

a)

60°

b)

36°

c)

80°

d)

54°

e)

110°

f)

25°

3 On a copy of the diagram, use ruler and compasses to construct the line from P perpendicular to the line segment AB.

× P

A ——————————————————————————— B

4 On a copy of the diagram, use ruler and compasses to construct the line from P perpendicular to the line segment AB.

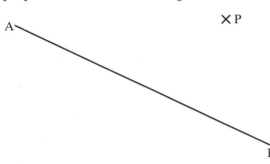

× P

A

B

5 On copies of the diagrams, use ruler and compasses to construct a line through P perpendicular to AB.

a)

A ———————————————— P ———— B

b)

A ———— P ———————————————— B

c)

A

P

B

6 a) Construct a triangle RST with sides
RT = 10 cm, RS = 13 cm, ST = 11 cm

b) (i) Construct the perpendicular
bisector of the side RS.
(ii) Construct the perpendicular
bisector of the side ST.
(iii) Construct the perpendicular
bisector of the side RT.
Your three bisectors should all meet at a
single point. Label this point X.

c) Draw a circle, centre X, that passes through R.
What do you notice?

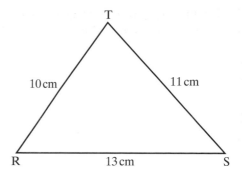

7 a) Construct a triangle EFG with sides
EF = 12 cm, FG = 14 cm, EG = 9 cm

b) (i) Bisect the angle EFG.
(ii) Bisect the angle FGE.
(iii) Bisect the angle GEF.
The three bisectors should
all meet at a single point.
Label this point Y.

c) Draw a circle, centre Y, that touches **one side**
of the triangle.
What do you notice?

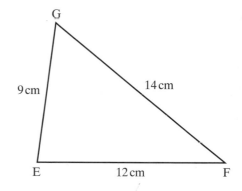

8 Follow this method to construct an angle of exactly 150° without a protractor.
Step 1   Draw a line 7 centimetres long.
Step 2   Use this line as one side of an equilateral triangle.
Use compasses and a straight edge to construct the other two sides.
You now have three angles of 60°
Step 3   Choose one of the angles and bisect it.
You now have an angle of 30°
Step 4   Extend one of the arms of the 30° angle.
The angle next to the 30° will be 150°

9 Follow this method to construct an angle of exactly 135° without a protractor.
Step 1   Draw a line 8 centimetres long.
Step 2   Construct the perpendicular bisector of your line.
You now have four angles of 90°
Step 3   Choose one of the angles and bisect it.
The 90° angle together with one of the 45° angles makes 135°

EXERCISE 18.3

In all your answers to this Exercise, you should leave all your construction lines showing.

1 Draw the locus of all the points which are exactly 5 cm away from the following:

a) •

b) ——————————————————————
    8 cm

2 a) Mark a point P on your paper.

b) Draw the locus of the points that are exactly 4.5 cm away from your point P.

c) Shade the region that contains all the points *more than* 4.5 cm away from P.

3 a) Mark two points, C and D, 7 cm apart.   C •————————————————————• D

b) Draw the locus of the points that are equidistant from C and D.

c) Shade the region that contains the points nearer to D than to C.

4 a) Mark two points, A and B, 10 cm apart on your paper.

   •   •
   A   B

b) Draw the locus of the points which are equidistant from A and B.

c) Shade the region that contains the points nearer to A than to B.

5 a) Mark two points, C and D, 8 cm apart on your paper.

b) Draw the locus of the points which are exactly 2.3 cm away from your line.

c) Shade the region which contains all the points *less than* 2.3 cm away from your line.

6 a) Use a protractor to draw two lines, PQ and PR at 50° to each other.

b) Draw the locus of the points that are equidistant from the lines PQ and PR.

c) Shade the region that contains the points nearer to PR than to PQ.

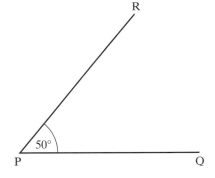

7 In the diagram, the rectangle represents a flowerbed. Poppy wants to make a path that is 2 m away from the edge of the flowerbed. Copy the diagram. Draw Poppy's path on your diagram.

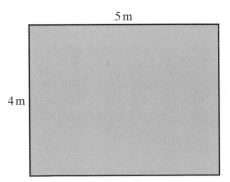

Scale 1 cm to 1 m

8 a) Use a protractor to draw two lines, AB and AC at 72° to each other.

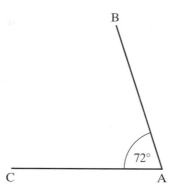

b) Draw the locus of the points which are equidistant from the lines AB and AC.
c) Shade the region which contains the points nearer to AB than to AC.

9 The vegetable patch in Josie's garden is rectangular in shape.

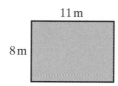

11 m

8 m

Josie wants to make a path around the *inside* of her vegetable patch.
She wants the path to be exactly 2 m wide and touching the perimeter of the vegetable patch.

a) Using a scale of 1 cm to 1 m, draw an accurate diagram of Josie's vegetable patch.
b) Accurately draw the path inside the vegetable patch.

10 In the diagram, the rectangle PQRS represents a farm.

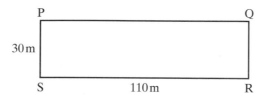

It is to be divided up into three regions.
    The region formed by all points within 30 metres of PS will be the farm house.
    The region formed by all points within 40 metres of Q will be the hen house.
    The rest of the farm will be grass.
Using a scale of 1 cm to 10 m, construct rectangle PQRS, and indicate each of the three regions.

11 The diagram shows a rectangular nature reserve.
A gravel path is to be laid. The centre line of the path runs diagonally from E to G.
All parts of the nature reserve within 1 metre of this centre line are to be gravelled.

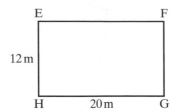

a) Using a scale of 1 cm to 2 m, construct rectangle EFGH, and mark the centre line of the path.

b) Shade the region to be gravelled.

A pond is to be built in the nature reserve.
The pond will occupy all points within 4.6 m of F.

c) Shade the area where the pond will be.

12 The diagram shows a paved play area at a school.

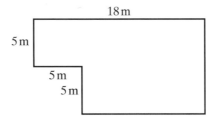

The local authority wants to erect a fence around the outside of the play area.
They decide that the fence should come to within 2 metres of the nearest point of the play area.
Using a scale of 1 cm to 2 m, make an accurate drawing to show the position of the fence.

13 The diagram shows a rectangular field ABCD.
Harry wants to bury some treasure in the field.

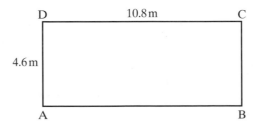

Harry buries the treasure exactly 3.5 m from DC, and exactly the same distance from AB and BC.
Using a scale of 1 cm to 1 m, construct rectangle ABCD and mark with a cross (✕) where Harry buried the treasure.

14 At the edge of a bay there
are three lighthouses.
The diagram shows the
position of the three
lighthouses, P, Q and R.

Draw points P, Q and R quite
far apart.

a) Using compasses and a
straight edge, construct
the locus of all points that
are equidistant from
P and Q.

b) Repeat the construction
using Q and R, and again,
using P and R.

c) Hence divide the bay into
three regions, one for each
lighthouse.
Use coloured pencils to mark the regions distinctly.

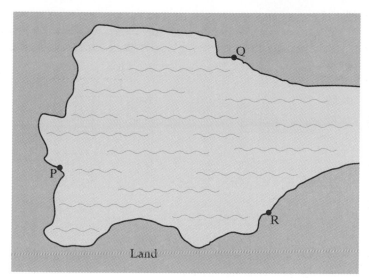

EXERCISE 18.4

1 Match together these 3-D solids with their nets.

A   B   C    D

1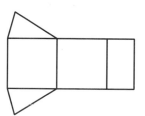

2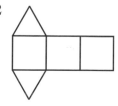

3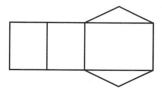

4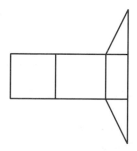

2 Which of these are nets of a cube?

A  B 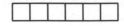 C

D 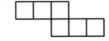 E F

3 On squared paper draw accurate nets of these solids:

a)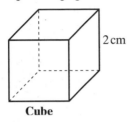

2 cm

Cube

b)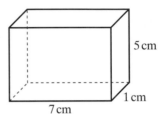

5 cm

1 cm

7 cm

Cuboid

c)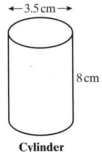

←3.5 cm→

8 cm

Cylinder
(circumference 11 cm)

d)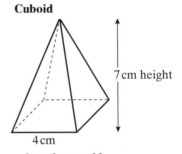

7 cm height

4 cm

Square-based pyramid

4 Sketch the 3-D shapes made by these nets.

a)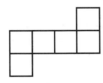

b)

c)

d)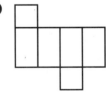

5 The isometric drawing shows some cubes forming a 3-D shape.

a) Copy the shape onto isometric paper.
b) On squared paper draw a sketch of
 (i) a plan view as seen from A
 (ii) a side elevation as seen from B
 (iii) a front elevation as seen from C.

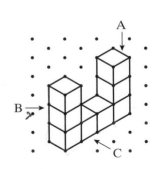

6 (i) Copy each of these 3-D shapes onto isometric paper.
  (ii) On squared paper, draw an accurate plan, front elevation (from A) and side elevation (from B) of each solid.
  (iii) Write down the volume of each shape.

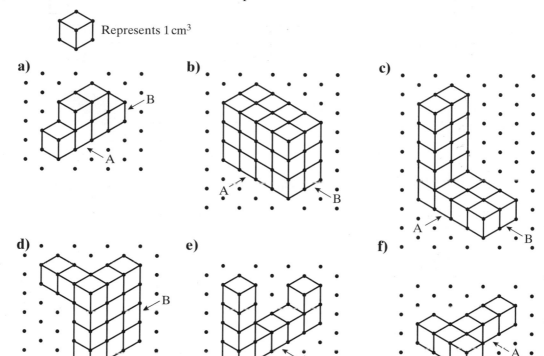

Represents 1 cm³

a)

b)

c)

d)

e)

f)

7 Here is a tin of sweets.
  The length of the label around the tin is 9 cm
  a) Draw an accurate net of the tin.
  b) Draw an accurate plan, front elevation and side elevations of the tin.

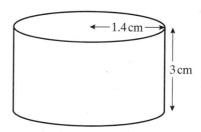

8 **(i)** Sketch each of these 3-D shapes onto isometric paper.
  (ii) On squared paper draw an accurate plan, front elevation and side elevation of each
    of these solids:

a)

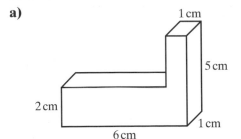

b)

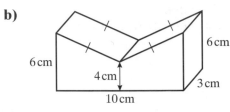

c)

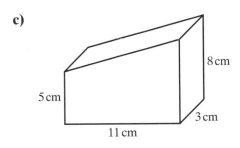

9 A pyramid has a square base whose sides are 5 cm long.
  The triangular faces have sides 7 cm, 7 cm and 5 cm.
  a) Sketch a plan view of the pyramid.
  b) Sketch a net for the pyramid.

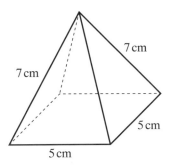

10 Kamini and Dakshesh have been building shapes with centimetre cubes on a
  square grid.
  The diagrams show plan views of their shapes.
  The numbers 1, 2, 3 tell you how many cubes are stacked on top of each square.

Kamini

3	1
1	2

A

Dakshesh

1	2
2	2

B

  a) Draw a front elevation to show how Kamini's shape appears seen from direction A.
  b) Make an isometric drawing of Dakshesh's shape, viewed from direction B.

18 Constructions and loci

11 Look at this cube.

a) How many:
  (i) faces
  (ii) edges
  (iii) vertices does it have?

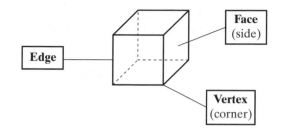

b) Copy and complete this table:

Solid	Number of faces	Number of edges	Number of vertices	Number of faces + 2	Number of edges minus number of vertices
A					
B					
C					
D					
E					
F					

c) What do you notice?

CHAPTER 19

Collecting data

EXERCISE 19.1

1 Fran carries out a survey.
She wants to see whether the school canteen should be open every day.
She decides to arrive at the canteen at 12 noon and ask the first 50 students who arrive the question:

'Don't you agree that our school canteen should stay open every day?'

a) Write down **two** reasons why this is **not** a good way to find out whether students at Fran's school think the school canteen should be open every day.
b) Describe a better method of selecting 50 students that Fran might use.
c) Write down a replacement question that Fran could use.

2 Keith is conducting a survey about restaurants in his town.
Here are three questions in his questionnaire.

1 How old are you?   _____

2 What restaurant do you go to?   _____

3 Do you agree that restaurants are too expensive?
   YES [ ]   NO [ ]   (Tick one)

a) Explain briefly why each of Keith's questions is unsuitable.
b) Design suitable replacements for each of Keith's questions.

3 Sarah is going to find out about the make-up habits of the female students and teachers at her school.
She has designed three questions:

1 Are you (Please tick)   Student [ ]   Teacher [ ]

2 How old are you?

3 How much make-up do you use?
   A lot [ ]   Not very much [ ]

a) Explain briefly why Sarah's questions 2 and 3 is unsuitable.
b) Design suitable replacements for questions 2 and 3 in Sarah's questionnaire.

4 Reeve wants to find out about the amount of pocket money students at her school get.
She has designed three questions.

1 Who gives you your pocket money?

2 Who is your favourite sports personality?

3 How much do you get?   £3–£5   £5–£7   £7 or more

a) Criticise each of Reeve's questions.
b) Design suitable replacements for questions 1 and 3 in Reeve's questionnaire.

5 Fred is carrying out a survey.
He wants to use a questionnaire to find out what pets students at his school like.
He also wants to see if boys like the same pets as girls.
Write down two questions that Fred might use in his questionnaire.

6 Design three suitable questions that Terri might use to test her hypothesis that boys spend more time playing sport than girls.

7 Bella and Victoria are going to interview 52 people at their youth club.
They want to choose the people in as fair a way as possible.

Explain carefully whether one of these methods is better than the other one.

EXERCISE 19.2

State whether the following sets of data would be categorical, discrete or continuous.

1 The number of cars in the school car park.

2 The shoe sizes of the people in your class.

3 The number of hours of television people watch each week.

4 The amount of water people drink in a day.

5 The weight of the contents in customers' shopping trolleys.

6 The heights of the dogs at a dog show.

7 The colours of people's clothes at a concert.

8 The amount waiters earn in tips at a restaurant.

9 The favourite chocolate bar of the students at your school.

10 The number of times a spinner lands on red.

11 The mark that the people in your class scored in a test.

12 The time students take to get to school.

13 The days of the week that people watch television.

14 The volume of water that people use in a bath.

15 The number of days people go abroad in a year.

EXERCISE 19.3

Do not use your calculator for Questions 1–4.

1   Kate carries out a survey about her friends' favourite pets.
    She makes a frequency table of her results.

Pet	Tally	Frequency
Cat	|||| |||| ||||	15
Dog	|||| |||	8
Hamster	|||| ||||		
Goldfish	||||	
Birds	||	
Rabbit	|||| |||| |||| |||	

a) Complete a copy of Kate's frequency table.
b) How many people liked goldfish best?
c) Which is the favourite pet of most of Kate's friends?
d) Which is the least favourite pet of most of Kate's friends?
e) How many friends did Kate ask altogether?

2   Anna collects data about the flavour of fruit juice people buy in the canteen.
    Here are her results.

lemon	mango	apple	lemon	orange	mango
apple	orange	orange	apple	mango	pear
apple	pear	mango	mango	lemon	apple
orange	mango	lemon	lemon	mango	mango
apple	mango	apple	apple	orange	mango

a) Complete a copy of the frequency table to show Anna's results.

Flavour of juice	Tally	Frequency
apple		
lemon		
mango		
orange		
pear		

b) Which flavour was the most popular?
c) How many more people preferred apple to pear?
d) Which two flavours were equally popular?

3 Lisa has carried out a survey to find out the marks students got in their last test.
Here are the marks for the students in her class.

45	55	73	28	87	67	49	52	57	40
20	57	47	92	78	45	61	25	70	42
52	43	41	22	74	74	45	33	62	51

a) Complete the frequency table to show Lisa's results.

Mark	Tally	Frequency
20 and under		
21–30		
31–40		
41–50		
51–60		
61–70		
71–80		
Over 80		

b) How many students got between 41 and 50 marks?
c) How many students scored more than 60?
d) How many students were in Lisa's class?

4 The weights of 36 potatoes were found.
The results, in grams, are shown below.

127	163	242	280	339	207	200	97	151	262	190	73
333	310	230	250	222	80	150	328	200	111	102	273
208	212	217	143	349	285	146	174	285	243	291	187

a) Complete the frequency table to show these results.

Weight, w (in g)	Tally	Frequency
$0 < w \leqslant 100$		
$100 < w \leqslant 150$		
$150 < w \leqslant 200$		
$200 < w \leqslant 250$		
$250 < w \leqslant 300$		
$300 < w \leqslant 350$		

b) How many potatoes weighed 200 grams or less?

5 Will collected some data on the number of hours his friends spent on the computer on Sunday.

Here are the times, correct to the nearest hour.

4	2	0	4	4	5	2	3	1	1	5	6
2	1	0	4	4	4	3	1	3	2	1	3

a) Copy and complete the frequency table.

Number of hours	Tally	Frequency
0		
1		
2		
3		
4		
5		
6		

b) How many of Will's friends spent 4 hours, to the nearest hour, on the computer?

c) How many of Will's friends spent more than 4 hours, to the nearest hour, on the computer?

d) How many hours did Will's friends spend on the computer in total?

EXERCISE 19.4

1 Jonathan has been carrying out a survey.
   He asked some tea and coffee drinkers whether they drank tea or coffee or both.
   Here are some of his results, transferred into a two-way table.

	Tea	Coffee	Both	Total
Male	15			42
Female		12	17	
Total			32	100

   a) Copy and complete the two-way table.   b) How many females drank only tea?
   c) How many men were in the survey?   d) How many people drank only coffee?

2 Jess asked 80 people to give their favourite colour from blue, green, red or yellow.
   This two-way table shows some of her results.

	Blue	Green	Red	Yellow	Total
Boys			8	6	
Girls	7			2	36
Total	20		31		80

   a) Copy and complete the two-way table.
   b) How many people liked red best?
   c) How many girls liked green best?
   d) How many boys were asked?

3 Jean has collected data on how much weight (*w*), in kg, people at her slimming club
   have lost over the last 6 months.
   This two-way table shows some of her results.

	Men	Women	Total
$0 < w \leqslant 5$	1		10
$5 < w \leqslant 10$		26	
$10 < w \leqslant 15$		17	27
over 15	9	11	
Total	27		

   a) Copy and complete the two-way table.
   b) How many women lost more than 15 kg?
   c) How many men lost more than 10 kg?
   d) How many women lost 10 kg or less?
   e) How many people took part in Jean's survey?

4 The two-way table shows the number of children and the number of pets in some families.

Number of children

		0	1	2	3 or more
Number of pets	**0**	4	4	3	1
	1	3	3	6	5
	2	2	3	4	2
	3 or more	5	1	0	0

a) How many families have 2 children and one pet?
b) How many families have 1 child and two pets?
c) How many families have 3 or more pets?
d) How many families have the same number of children as pets?
e) How many families took part in the survey?

5 The two-way table shows the number of cinemas and the number of large supermarkets in some small towns.

Number of supermarkets

		1	2	3	4
Number of cinemas	**0**	0	7	8	3
	1	2	8	6	4
	2	1	10	15	2
	3	0	6	7	1

a) How many small towns have 4 supermarkets and 2 cinemas?
b) How many small towns have exactly 1 supermarket?
c) How many small towns have the same number of supermarkets as cinemas?
d) How many small towns have fewer supermarkets than cinemas?

6 Mel is carrying out a survey to see whether teachers and students like the same music.
She asks each person in the sample to name their favourite type of music.
Mel records the results on a data collection sheet.
Here are her results after asking the first 14 people.

Pop	Classical	Jazz	Folk	Teachers	Students
⦀⦀ || ||	⦀⦀			⦀⦀		⦀⦀ ⦀⦀

a) Explain one disadvantage with recording the data in this way.
b) Design an improved data collection sheet that Mel might use.

7 Design a data collection sheet for a survey about how much homework students get.

CHAPTER 20

Working with statistics

EXERCISE 20.1

Do not use your calculator for Questions 1–3.

1 Find: **(i)** the mean, **(ii)** the median and **(iii)** the mode of the following sets of data.

a)	4	4	5	5	8	8	8	
b)	5	7	7	7	11	12	13	14	14
c)	10	10	11	13	16			

2 Pat surveys her friends to find out about their favourite colours.
Here are her results.

red	blue	green	blue	black
green	red	yellow	white	orange
pink	purple	brown	red	black

Which colour is the mode?

3 Mina surveys some students about their favourite subject at school.
Here are her results.

Subject	Frequency
Art	25
Drama	38
English	18
Mathematics	4
Science	2

Which subject is the mode?

You can use a calculator for Questions 4–7.

4 Find: **(i)** the mean, **(ii)** the median and **(iii)** the mode of the following sets of data.

a)	5	2	9	8	0	2	3	7	1	
b)	64	68	75	63	68	74	72	63	78	70	73
c)	125	136	128	130	130	129	120			

5 Find: **(i)** the mean, **(ii)** the median and **(iii)** the mode of the following sets of data.

a)	4	7	9	12	4	2	13	10	
b)	30	38	48	41	25	27	41	38	41	32
c)	43	35	50	31	37	43			

6 Jeremy wants to sell his business.
He puts an advertisement in the local paper.

> **BUSINESS FOR SALE**
> Average annual profit
> £80 000
> Telephone: 8488 5901

Here are his annual profits (to the nearest £10 000) for the past 15 years.

£200 000	£180 000	£150 000	£160 000	£170 000
£140 000	£90 000	£80 000	£60 000	£50 000
£50 000	£40 000	£50 000	£50 000	£30 000

a) Find the mean profit.
b) Find the median profit.
c) Which average has the advertisement used?
Is this a fair average to take?
Give a reason for your answer.

7 Joe has recorded the temperature at midnight for 12 nights during February.
Here are his results.

2 °C −5 °C −4 °C 0 °C 1 °C −3 °C
−2 °C 1 °C 2 °C 1 °C −2 °C −5 °C

Find the:
a) mean temperature   **b)** median temperature   **c)** modal temperature.

EXERCISE 20.2

1. Find: **(i)** the mean, **(ii)** the median, **(iii)** the mode and **(iv)** the range of the following sets of data.

a)	3	7	3	7	7	1	1	7	6	2
b)	13	17	13	17	17	11	11	17	16	12
c)	73	77	73	77	77	71	71	77	76	72
d)	213	217	213	217	217	211	211	217	216	212

e) What do you notice about your answers?

You can use a calculator for Questions 2 and 3.

2 The data set below gives the total number of marks scored by each of 21 students in a test.

67	54	39	63	54	70	43
60	50	64	38	61	87	68
42	55	66	80	33	41	35

a) Work out the range of the data.
b) Write down the mode.
c) Calculate the value of the mean.
d) Find the value of the median.

3 Here are the heights and the weights of 8 students.

Name	Sarah	Rose	Anna	Lisa	Mia	Alison	Jo	Pat
Height	1.5 m	1.27 m	1.42 m	1.67 m	1.27 m	1.4 m	1.61 m	1.56 m
Weight	78.2 kg	95 kg	73.7 kg	85 kg	62.1 kg	80.2 kg	85 kg	62 kg

a) Work out the:
    (i) mean   **(ii)** median   **(iii)** range   for the heights.
b) Work out the:
    (i) mean   **(ii)** median   **(iii)** range   for the weights.
c) Which person was not very tall but weighed a lot?

EXERCISE 20.3

1 A company sells beds in flat packs for customers to build.
This stem and leaf diagram shows the times, in minutes, it took for each of 14 people to build a bed.

```
2 | 4 5
3 | 3 8 8
4 | 0 1 7 9
5 | 2 5 6
6 | 1 7
```
Key: 2 | 4 = 24

Four more people build a bed from a flat pack.
Their times, in minutes are 52, 48, 69 and 52
a) Draw a new stem and leaf diagram to include the four new people.
b) Find the mode.
c) Find the median.
d) Find the range.

2 Patsy has made an unsorted stem and leaf diagram to show the weights, in grams, of 23 loaves of bread in her bakery.

```
70 | 6 4 1
71 | 7 4 0 6 7
72 | 3 9 5 4
73 | 9 8 5 8 3 8
74 | 6 6 4 9 0
```
Key: 70 | 6 = 706

a) Redraw the diagram so that it is fully sorted.
b) Find the mode.
c) Find the median.
d) Find the range.

3 There were 15 pairs of trainers in a sale.
Their prices, in £, are shown below:

23	38	29	53	46	40	23	37
20	34	56	42	33	39	21	

a) Draw a stem and leaf diagram to show this information.
Remember to include a key.
b) Work out the range of the prices.
c) Find the median price.

4 A cycling group went on a 13-day holiday.
Here are the number of kilometres they planned to cycle each day.

18	32	23	28	19	40	41
25	34	38	19	22	27	

a) Draw a stem and leaf diagram to show this information.
(Don't forget the key.)
b) Work out the range.
c) Find the median.

5 16 students sat an English examination.
Here are their marks:

72	68	53	60	72	50	83	57
50	75	81	76	65	69	60	71

a) Draw a stem and leaf diagram to show this information.
b) Find the median of the marks.

EXERCISE 20.4

1 Alan is an estate agent.
He gathers some data on the number of apartments available in 100 large apartment blocks.

Number of apartments	Frequency
1	28
2	35
3	21
4	9
5	7
Total	100

a) What is the modal number of apartments available?
b) Find the median number of apartments available.
c) Work out the mean number of apartments available.
d) What is the range of Alan's data?

You can use a calculator for Questions 2 and 3.

2 Lilian conducts a survey.
She finds out how many hours of TV students watched last Monday.
The frequency table shows her results.

a) How many students took part in Lilian's survey?

The average number of hours of TV watched is 4.

Lilian

Number of hours watching TV	Frequency
0	2
1	7
2	9
3	4
4	3
5 or more	0
Total	

b) Explain why Lilian must be wrong.
c) Find the modal number of hours of TV watched.
d) Find the range of the data.
e) Work out the mean number of hours of TV watched.
f) Find the median number of hours of TV watched.

3 Jasmine is surveying the local hockey club.
She records the number of goals scored in each hockey match in the past three years.

a) How many matches did Jasmine survey?
b) What is the modal number of goals per match?
c) Find the median number of goals per match.
d) Work out the mean number of goals per match.
e) What is the range of Jasmine's data?

Number of goals	Frequency
0	3
1	2
2	5
3	7
4	2
5	3
6	0
7	0
8	1
Total	

EXERCISE 20.5

1 Hajra has measured the weights of some babies and toddlers at a crèche.
The frequency table shows her results.

Weight (w) in kg	Frequency (f)
$0 < w \leqslant 4$	5
$4 < w \leqslant 8$	6
$8 < w \leqslant 12$	7
$12 < w \leqslant 16$	2
$16 < w \leqslant 20$	1
Total	

a) State the modal class.
b) Find the class interval in which the median lies.
c) Estimate the mean weight of these babies and toddlers.
Show all your working clearly.

You can use a calculator for Questions 2–4.

2 Mario records the number of people at the gym each day in April.
The frequency table shows his results.

Number of people at the gym	Frequency	Midpoint
28 to 34	4	
35 to 41	8	
42 to 48	12	
49 to 55	6	
Total		

a) Copy the table, and fill in the midpoint values in the third column.
b) Use the table to help you calculate an estimate of the mean number of people at the gym each day in April.
c) State the modal class.
d) Find the class interval which contains the median.
e) Carlo says, 'The range is 30.'
Explain why Carlo must be wrong.

3 The frequency table shows the ages of the passengers on a cruise liner on the Atlantic.

a) State the modal class.
b) Find the class interval which contains the median.
c) Work out an estimate of the mean age of the passengers on the cruise liner. Give your answer correct to three significant figures.

Age (A) in years	Frequency
$0 < A \leqslant 20$	38
$20 < A \leqslant 40$	70
$40 < A \leqslant 60$	75
$60 < A \leqslant 80$	15
$80 < A \leqslant 100$	2
Total	

4 Seamus solves some Sudoku puzzles.
The frequency table shows the times taken to solve each puzzle.

Time (t) in minutes	Frequency
$5 \leqslant t < 10$	4
$10 \leqslant t < 15$	5
$15 \leqslant t < 20$	7
$20 \leqslant t < 25$	9
$25 \leqslant t < 30$	5
Total	

a) State the modal class.
b) Find the class interval which contains the median.
c) Calculate an estimate of the mean time Seamus took to solve the puzzles.
d) Explain briefly why your answer can only be an estimate.

EXERCISE 20.6

1 Match together these graphs with their descriptions.

P

Q

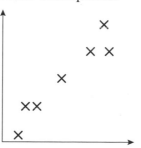

R

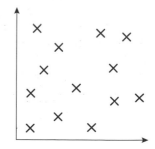

A   Length of hair against Height
B   Value of car against Age of car
C   Height against Weight

2 A park has an outdoor swimming pool.
The scatter graph shows the maximum temperature and the number of people who used the pool on ten Saturdays in summer.

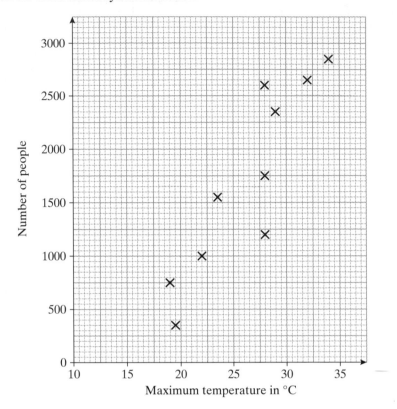

a) Describe the correlation between the maximum temperature and the number of people who used the pool.

b) On a copy of the scatter graph, draw a line of best fit.

The weather forecast for the next Saturday gives a maximum temperature of 27 °C

c) Use your line of best fit to estimate the number of people who will use the pool.

[Edexcel]

3 Some students took a mathematics test and a science test.
The scatter graph shows information about the test marks of eight students.

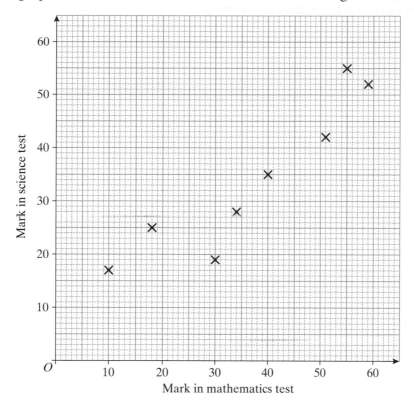

The table shows the test marks of four more students.

Mark in mathematics test	14	25	50	58
Mark in science test	21	23	38	51

a) On a copy of the scatter graph, plot the information from the table.
b) Draw a line of best fit on the scatter graph.
c) Describe the **correlation** between the marks in the mathematics test and the marks in the science test.

[Edexcel]

4 Ten men took part in a long jump competition.
The table shows the heights of the ten men and the best jumps they made.

Best jump (m)	5.33	6.00	5.00	5.95	4.80	5.72	4.60	5.80	4.40	5.04
Height of men (m)	1.70	1.80	1.65	1.75	1.65	1.74	1.60	1.75	1.60	1.67

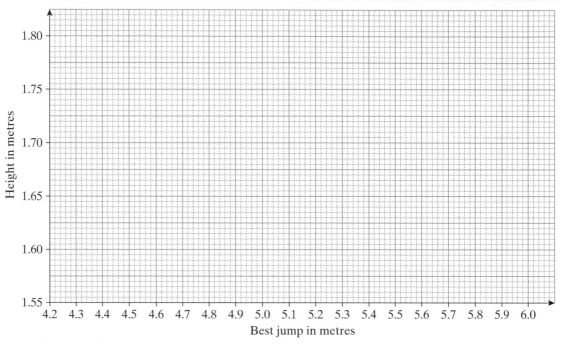

a) On a copy of the grid above plot the points as a scatter diagram.
b) Describe the relationship between the height and the maximum difference.
c) Draw in a line of best fit.

[Edexcel]

5 Here is a scatter graph.
One axis is labelled 'Height'.
a) For this graph, state the type
of correlation.
b) From the list below, choose the most
appropriate label for the other axis:

Length of hair
Number of sisters
Length of legs
GCSE French mark

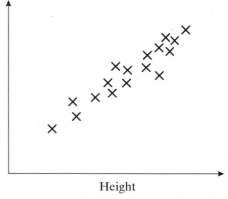

[Edexcel]

CHAPTER 21

Presenting data

EXERCISE 21.1

1 Simone wrote down the favourite colours of all the students in her class.
The bar chart shows her results.

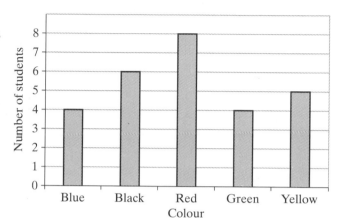

a) Which colour was the favourite for five of the students?

b) Which was the favourite colour?

c) Two colours were equally popular.
Which colours were they?

d) How many students were in Simone's class?

2 A superstore uses a pictogram to show the number of DVD players they sold over a 4-week period.

a) How many DVD players were sold in week 1?

b) In which week were the least number of DVD players sold?

c) How many DVD players were sold in week 3?

d) How many DVD players were sold in week 4?

The superstore has data for two more weeks.

Week number	Number of DVD players sold
Week 5	16
Week 6	9

e) Add the data to a copy of the pictogram.

3 Fatima recorded the maximum temperatures in London and in Cape Town for the first six months of last year.
She produced a bar chart to compare the temperatures in the two cities.

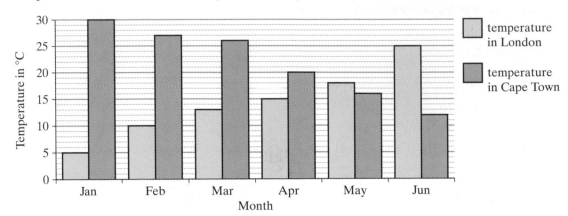

Use the dual bar chart to write down the following.
   a) the maximum temperature in February in:
       (i) London   **(ii)** Cape Town
   b) the month which had the highest maximum temperature in:
       (i) London   **(ii)** Cape Town
   c) the month where the difference in maximum temperature between the two cities was:
       (i) the highest   **(ii)** the lowest
       and write down what this difference was.
   d) the month where the highest maximum temperature in London became greater than the highest maximum temperature in Cape Town.

4 An art gallery had a sale of paintings.
The owner wrote down her total sales on each of the first five days of the sale.
The information is in the table below.

Day of the sale	Total sales
Monday	£4000
Tuesday	£3000
Wednesday	£1500
Thursday	£2250
Friday	£3750

Draw a pictogram to show this information.

Use ⊕ to represent £1000

5 Six students, Ann, Ben, Con, Dani, Ed and Flo, each sat an examination in English, Mathematics and Science.
Each examination was marked out of 100
Their teacher made a bar chart of their results.

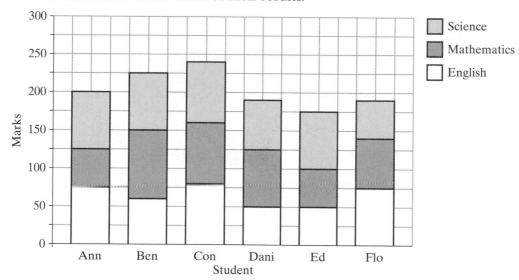

a) Which student got the highest total mark?
b) Which student got the highest Mathematics mark?
c) Two students got the same overall total.
Write down the names of these two students.
d) One student got the same mark for all three subjects.
Write down the name of this student.
e) What did Ann score for Science?
f) One student's best mark was for English.
Write down the name of this student.

6 May counts the number of each of her favourite animals at the zoo.
Here are her results.

Type of animal	Number of animals
Giraffes	3
Lions	6
Monkeys	10
Polar Bears	3
Zebras	8

Draw a bar chart to show this information.

EXERCISE 21.2

1 Des asked 40 of his friends which was their favourite holiday destination in the UK. The table shows the results.

Draw a pie chart to illustrate this information.

City	Frequency
Blackpool	5
Bournemouth	7
Brighton	10
Falmouth	2
London	16

2 An optometrist wrote down what he sold to the 18 patients he saw on Saturday.

Draw a pie chart to illustrate this information.

Sales	Number
Spectacles	12
Contact lenses	4
Nothing	2

3 Patsy earns £30 a week doing various jobs. The table shows how she spent this money last week.

Draw a pie chart to illustrate this information.

Cinema	£7
Clothes	£12
Eating out	£8
Magazines	£3

4 Richard has 60 books on the bookshelves in his room.

Draw a pie chart to illustrate this data.

Type of book	Number of books
Science Fiction	17
Adventure	7
Reference	15
Novel	21

5 The table shows the spread of the ages of the members of a sports club.

Draw a pie chart to illustrate this data.

Age of members	Percentage
under 18	10
$18 \leqslant age < 25$	18
$25 \leqslant age < 40$	42
$40 \leqslant age < 60$	22
60 or over	8

6 Craig draws a pie chart to show how he uses his earnings in a typical month.

a) Which of his spending is the mode?

Craig earns £720 per month after tax.

b) Work out how much Craig spends on:
   (i) household bills
   (ii) entertainment
   (iii) rent
   (iv) food

c) How much does Craig save each month?

7 Janet and John each draw a pie chart to show how long they spent on their homework last weekend.

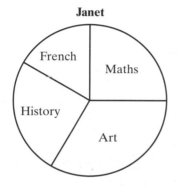

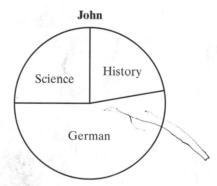

Janet spends 6 hours on her homework.
John spends 8 hours on his homework.

a) Write down the subject which is the mode for the:
   (i) Janet   **(ii)** John

b) Work out how much time Janet spends on:
   (i) Maths   **(ii)** Art

c) Work out how much time John spends on Science.

d) Max says that Janet spends longer on her history homework than John because the angle for history on her pie chart is larger than the angle for history on John's pie chart. Explain why Max is wrong.

EXERCISE 21.3

Do not use your calculator for Questions 1 and 2.

1 Here are the ages of the people staying in a hotel.
  a) Draw a frequency diagram to show this data.
  b) How many people were staying in the hotel?
  c) Write down the modal class interval.

Age	Frequency
$0 \leqslant$ age < 15	15
$15 \leqslant$ age < 30	11
$30 \leqslant$ age < 45	47
$45 \leqslant$ age < 60	32
$60 \leqslant$ age < 75	10
$75 \leqslant$ age < 90	4

2 Lara carries out a survey to find out the average amount of time students spend watching TV at the weekend.
Here are her results.
Draw a histogram of Lara's results.

Time in hours (h)	Frequency
$0 \leqslant h < 3$	5
$3 \leqslant h < 6$	14
$6 \leqslant h < 9$	22
$9 \leqslant h < 12$	8

You can use your calculator for Questions 3 and 4.

3 Maddy carried out a survey of the time, in minutes, it took students to get to school. The table shows the results of her survey.
  a) Draw a histogram to show this data.
  b) Write down the modal class interval.
  c) How many students were in Maddy's survey?

Time, t, in minutes	Frequency
$0 \leqslant t < 10$	11
$10 \leqslant t < 20$	34
$20 \leqslant t < 30$	26
$30 \leqslant t < 40$	15
$40 \leqslant t < 50$	4

  d) How many students took less than half an hour to get to school?
  e) Find an estimate for the mean time, in minutes, it took students to get to school.

4 A vet recorded the weight of 60 animals he saw over the weekend.
  a) Draw a histogram to show this data.
  b) Write down the modal class interval.
  c) Estimate the mean weight of the animals.
  d) Write down the class interval which contains the median.

Weight, w, in kg	Frequency
$0 \leqslant w < 3$	14
$3 \leqslant w < 6$	8
$6 \leqslant w < 9$	10
$9 \leqslant w < 12$	12
$12 \leqslant w < 15$	16

EXERCISE 21.4

1 Draw a frequency polygon for each of the sets of data given in Exercise 21.3.

You can use your calculator for Questions 2 and 3.

2 A football club carried out a survey of the ages of its 274 supporters at a football match. The results of the survey are shown in the table.

Age group, x, years	Frequency
$0 \leqslant x < 10$	24
$10 \leqslant x < 20$	36
$20 \leqslant x < 30$	60
$30 \leqslant x < 40$	52
$40 \leqslant x < 50$	30
$50 \leqslant x < 60$	44
$60 \leqslant x < 70$	28

a) Draw a frequency polygon to show this information.
b) Which is the modal class?
c) Find the class interval which contains the median.
d) Find an estimate for the mean age.

3 Justin is a florist.
He measured 100 flowers and recorded their heights to the nearest cm.
The table shows information about the heights, h, of the 100 flowers.

Height, h	Frequency, f
$0 \leqslant h < 20$	24
$20 \leqslant h < 40$	38
$40 \leqslant h < 60$	16
$60 \leqslant h < 80$	12
$80 \leqslant h < 100$	10

a) Draw a frequency polygon to show this information.
b) Work out an estimate for the mean height of these flowers.
c) Find the class interval that contains the median.

1 Lauren spent a year travelling.
She had a flight ticket to fly to
12 different countries in any order.
Each month she went to a different
country and recorded the
temperature there, in °C, on the
10th day of each month.
The graph shows this information.

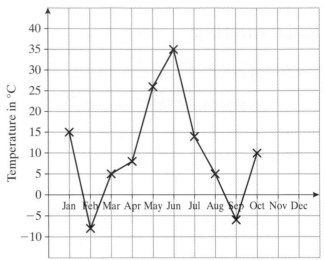

a) What is the range of the
temperatures?

b) Write down the temperature in:
(i) July, **(ii)** September

c) Lauren takes two more readings:

Month	Nov	Dec
Temp	−5 °C	27 °C

Add these measurements to a copy of Lauren's graph.

2 The time-series graph shows the
takings of a small company
one year.

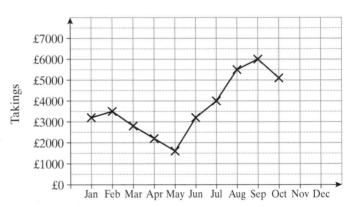

a) What is the range of the
takings?

b) Which two months have the
same takings?

c) Here is the data for November
and December.

Month	Nov	Dec
Takings	£4500	£5700

Add the data to a copy of the time-series graph.

3 Mr Smith recorded the number of customers in his shop at particular times one Friday.
Here are his results.

Time	9 am	10 am	11 am	12 noon	1 pm	2 pm	3 pm	4 pm	5 pm
No. customers	3	5	6	5	8	9	6	4	2

Draw a time-series graph for this data.

CHAPTER 22

Probability

EXERCISE 22.1

1 The colours on a regular 5-sided spinner
are red, blue, yellow, green and purple.
Dean spins the spinner once.
Find the probability that it
  a) lands on red
  b) does **not** land on green
  c) lands on black.

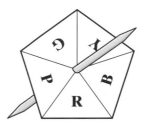

2 A football match is either won, lost or drawn.
On a copy of the probability line below mark with a letter **W** the probability that a
football match will be won.

3 A fair dice has sides numbered 1, 2, 3, 4, 5 and 6
The dice is thrown once.
On a copy of the probability line below:

  a) mark with a letter **A** the probability that the dice will land on an **even** number
  b) mark with a letter **B** the probability that the dice will land on a number **bigger** than 4
  c) mark with a letter **C** the probability that the dice will **not** land on a number bigger
than 4
  d) mark with a letter **D** the probability that the dice will land on a number **bigger** than 0
  e) mark with a letter **E** the probability that the dice will land on a number **smaller** than 1

4 An eight-sided spinner has sides 1, 2, 3, 4, 5, 6, 7 and 8
The spinner is spun once.
On a copy of the probability line below:

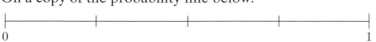

  a) mark with a letter **A** the probability that the spinner will
land on an odd number
  b) mark with a letter **B** the probability that the spinner will land on a 5
  c) mark with a letter **C** the probability that the spinner will land on a number less than 10
  d) mark with a letter **D** the probability that the spinner will land on a factor of 15
  e) mark with a letter **E** the probability that the spinner will **not** land on a number less
than 4
  f) mark with a letter **F** the probability that the spinner will land on a number bigger
than 2

5 The probability that Deborah wins a competition is 0.27
What is the probability that Deborah does not win the competition?

6 A bag contains 27 balls.
There are 10 blue balls and 13 red balls.
The rest of the balls are black.
A ball is chosen at random.
Find the probability that this ball is
a) red   **b)** **not** blue   **c)** black.

7 The probability that I will be at school tomorrow is 0.93
Work out the probability that I will **not** be at school tomorrow.

8 A drawer contains 40 socks.
10 of them are black, the rest are white or grey.
There are four times as many white socks as grey socks.
A sock is chosen at random.
a) Work out the probability that it is black.
b) Work out the probability that it is **not** black.
c) Work out the probability that it is white.

EXERCISE 22.2

Do not use your calculator for Questions 1 and 2.

1 The diagram on the right shows some shapes.
a) Copy and complete the table to show the
number of shapes in each category.

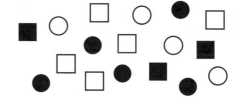

	Black	**White**	**Total**
Square			
Circle			
Total			

One of the shapes is chosen at random.
b) Write down the probability that the shape is a circle.
c) Write down the probability that the shape is **not** black.
d) Write down the probability that the shape is a black square.

2 The two-way table shows some information about the numbers of people in a library.
a) Copy and complete the two-way table.
A person in the library chooses a book.
b) Work out the probability the
person is a child.
One of the people in the library uses
the computer.
c) Work out the probability that the
person is male.

	Adults	**Children**	**Total**
Male		7	18
Female	20		
Total			50

You can use your calculator for Questions 3 and 4.

3 A small play group has children aged 2 years, 3 years and 4 years. Darren wrote down the ages of the children that attended on Tuesday.
The two-way table shows the results.

	2 years	3 years	4 years	Total
Boys	5	9	5	19
Girls	6	5	7	18
Total	11	14	12	37

    a) Find the probability that a child is a 2-year-old girl.
    b) Find the probability that a 4-year-old is a boy.
    c) Find the probability that a child is a 3-year-old.

4 A restaurant serves fruit salad, cheesecake or mousse for dessert.
Mr Chen keeps a record of what he serves on Sunday.
Every customer had one dessert.
The table shows some of this information.

	Fruit Salad	Cheesecake	Mousse	Total
Midday	10			
Evening		8		46
Total	28		25	70

    a) Copy and complete the two-way table.
    b) Find the probability that a customer had cheesecake for dessert.
    c) Find the probability that a customer had mousse for dessert in the evening.

EXERCISE 22.3

1 Two fair spinners are spun together. One spinner has five sides and the other has seven sides. The scores are **added** together.
    a) Copy and complete the table to show all the possible outcomes.
    b) What is the total number of possible outcomes?
    c) Write down the probability of getting a total of:
        (i) 9   **(ii)** 6
        (iii) 11   **(iv)** **not** 7
        (v) an even number   **(vi)** 10 or more

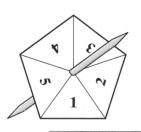
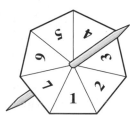

+	1	2	3	4	5	6	7
1	2						
2				6			
3		5					10
4							
5				9			12

2 A fair coin and a fair five-sided spinner
are thrown together.
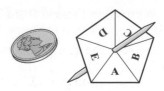
   a) Write down all the possible outcomes of throwing
   the fair coin and the fair spinner.
   b) Write down the probability of getting:
      (i) a 'head' and a 'D'   **(ii)** a 'tail' and **not** a C.

3 Two fair spinners are spun together.
One spinner has five sides and the
other has eight sides.
The scores are **multiplied** together.

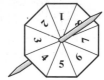
   a) Copy and complete the table to
   show all the possible outcomes.
   b) What is the total number of possible
   outcomes?
   c) Write down the probability of
   getting a total of:
      (i)   12
      (ii) 130
      (iii) 24
      (iv) **not** 18
      (v)   an odd number
      (vi) 25 or more.

×	1	2	3	4	5	6	7	8
1	1							
2						12		
3		6					21	
4					20			
5								40

4 Gloria eats dinner in a restaurant.
She randomly selects a starter, a main
course and a dessert from this list.
   a) List all the possible outcomes of
   randomly selecting a starter, a
   main course and a dessert.
   b) Write down the probability that Gloria will eat:
      i) pâté   **(ii)** soup and chicken   **(iii)** pâté, lamb and cherry pie.

Starter	Main course	Dessert
Soup	Chicken	Cherry pie
Pâté	Lamb	Fruit salad

5 Two fair spinners are spun together.
One spinner has four sides and the other has
six sides.
The **difference** between the two scores is found.

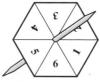
   a) Complete the table to show all the
   possible outcomes.
   b) What is the total number of
   possible outcomes?
   c) Write down the probability of
   getting a total of:
      (i)   1
      (ii) 5
      (iii) 2
      (iv) **not** 0
      (v)   an even number
      (vi) 4 or less

Difference	1	2	3	4	5	6
1	0					
2			1			
3				2		
4	3					

6 Liz picks one shape from Box P.
She then picks one shape from Box Q.

One pair she could pick is (white rectangle, white circle).

P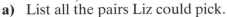

Q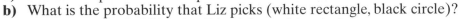

a) List all the pairs Liz could pick.
b) What is the probability that Liz picks (white rectangle, black circle)?
c) What is the probability that Liz picks a white circle?
d) What is the probability that Liz picks two circles?

7 Geoff picks one number from each of Boxes A, B and C.
One trio he could pick is (2, 4, 1)

A	B	C
9	4	5 7
2	8	1

a) List all the trios Geoff could pick.
b) Write down the probability that:
   (i) Geoff picks (2, 4, 1)   **(ii)** Geoff picks a '7'   **(iii)** Geoff picks an '8'
   (iv) Geoff picks a '9' and a '5'   **(v)** Geoff picks three numbers that total 24

EXERCISE 22.4

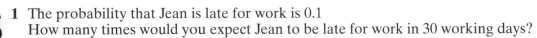

Do not use your calculator for Questions 1–4.

1 The probability that Jean is late for work is 0.1
How many times would you expect Jean to be late for work in 30 working days?

2 A fair coin is thrown 300 times.
How many times would you expect it to show 'heads'?

3 Simon has a biased dice.
Simon tosses the dice 100 times.
Here are his results.

Result	1	2	3	4	5	6
Frequency	42	27	2	3	21	5

Write down the experimental probability of the dice landing on:
a) 5   **b)** 1   **c)** an odd number

4 The probability of a biased coin showing 'heads' is 0.25
a) What is the probability of the coin showing 'tails'?

The coin is thrown 40 times.
b) Work out an estimate for the number of times it will show 'tails'.

You can use your calculator for Questions 5–8.

5   The probability that Colin will put up his hand in a lesson is $\frac{4}{5}$.
   a)   What is the probability that Colin will not put up his hand in a lesson?
   Colin goes to 40 lessons each week.
   b)   Work out an estimate for the number of times Colin will put up his hand in a week.

6   There are 500 sweets in a jar. 100 of the sweets are red.
   The rest are orange, yellow or green.
   Ted takes a sweet at random from the jar and puts it back.
   a)   Calculate the probability that Ted takes a red sweet.
   Tessa takes 50 sweets at random from the jar.
   20 of these sweets are orange.
   b)   Estimate how many orange sweets are in the jar.

7   A box contains just 1p coins, 2p coins and 5p coins.
   There are 200 coins in the box and 50 of these are 2p coins.
   Miriam randomly takes a coin from the box and then puts it back.
   a)   Work out the probability that the coin Miriam takes is a 2p coin.
   Miriam then randomly takes 40 coins from the box. 10 of these were 5p coins.
   b)   Estimate the total number of 5p coins in the box.

8   In the sixth form at Gravitas School there are 63 boys and 78 girls.
   a)   A student is chosen at random from the sixth form.
          Work out the probability that the student is a girl.
   b)   There are 1200 students altogether at Gravitas School.
          Estimate the total number of girls at the school.
   c)   Explain why your estimate might not be very reliable.

EXERCISE 22.5

1   When Lara returns from school she either does her homework or watches TV or tidies
   her room or has a snack.
   The probability that Lara does her homework is 0.2
   The probability that Lara watches TV is 0.4
   The probability that Lara has a snack is 0.3
   What is the probability that Lara tidies her room?

2   Joan either has coffee or tea or hot chocolate in the morning.
   The probability that Joan has coffee is 0.4
   The probability that Joan has tea is 0.25
   What is the probability that Joan has hot chocolate in the morning?

3   Traffic lights show either red or amber or green.
   The probability that the traffic lights show red is 0.43
   The probability that the traffic lights show green is 0.46
   a)   What is the probability that the traffic lights show amber?
   Janet uses 20 traffic lights on her way to work.
   b)   Work out an estimate for the number of times the traffic lights will show green.

4 A bag contains black, white, pink and orange counters.
A counter is drawn at random.
The table shows the probabilities of the counter being each of the four colours.

Colour	Black	White	Pink	Orange
Probability	$\frac{8}{23}$		$\frac{3}{23}$	$\frac{2}{23}$

a) What is the probability of the counter being white?
b) What is the probability of the counter being black or pink?
c) What is the probability of the counter **not** being orange?

5 A biased dice shows scores of 1, 2, 3, 4, 5, 6 with these probabilities.

Score	1	2	3	4	5	6
Probability	0.23		0.3	0.18	0.1	0.05

The dice is rolled once. Find the probability that the score obtained is:
a) 2   **b)** not 4   **c)** 5 or 6   **d)** an odd number.

6 When Andrew goes to work he is either early, on time or late.
The incomplete table shows some probabilities.

	Early	On time	Late
Probability	$\frac{27}{100}$		$\frac{6}{100}$

a) What is the probability that Andrew arrives on time?
b) Which of the three outcomes is most likely?
c) Find the probability that Andrew is **not** late for work.

7 There are four types of animals I might see in my garden.
The table shows the probability that a fox or a hedgehog or a squirrel or a
cat will be seen in my garden. No two animals are ever in the garden at the same time.

Type of creature	Fox	Hedgehog	Squirrel	Cat
Probability	0.15		0.4	0.28

a) Copy and complete the probability table.
b) Which type of animal is the most common in my garden?

An animal is observed at random.
Find the probability that:
c) it is **not** a cat,
d) it is a fox or a squirrel.

8 Bertha has three children, Robert, Johnny and Stephen.
Every Friday night Bertha has dinner with one of her children.
The probability that she has dinner with Robert is 0.28 and the probability
that she has dinner with Johnny is 0.3
a) Work out the probability that she has dinner with Stephen.
b) Work out the probability that she does **not** have dinner with Stephen.

9 A biased spinner gives scores of 1, 2, 3 or 4
   The probability of getting 1 is 0.37
   The probability of getting 2 is 0.18
   The probability of getting 4 is 0.2
   a) Calculate the probability of getting an even score.
   b) Work out the probability of getting a score of 3

10 Sue often partners Janet or Don in tennis.
   The probability that she will partner Janet is 0.38
   The probability that she will partner Don is 0.17
   a) Work out the probability that she will partner Janet or Don in tennis.
   b) Work out the probability that she will **not** partner either of these two in tennis.

11 Alex goes to the cinema.
   He can choose from a comedy, a drama, a science fiction movie or a horror movie.
   The table shows the probability that Alex will choose a particular type of movie.

Type of movie	Comedy	Drama	Science fiction	Horror
Probability	0.17	0.23	x	x

   a) Find the probability that he chooses a comedy or a drama.
   b) Find the probability that he chooses a science fiction movie or a horror movie.

   The probability he will choose to see a science fiction movie is the same as the
   probability he will choose to see a horror movie.
   c) Find the probability he will choose to see a horror movie.

12 The junior football team at Roddington School is picked from students in Years 7
   or 8 or 9
   Mr Right selects the junior football teams. The probability that he might select a
   student from Years 7 or 8 is shown in the table below.

Year group	Year 7	Year 8	Year 9
Probability	24%	32%	

   a) Copy and complete the probability table.
   Mr Right needs to pick 50 students for some junior football teams.
   b) Estimate the number of Year 8 students he is likely to pick.

13 A drawer contains black, white, blue and brown socks.
   In a probability experiment, one sock is chosen at random, and removed from the drawer.
   Its colour is noted, and it is returned to the drawer.
   The table shows some probabilities for this experiment.

Colour	Black	White	Blue	Brown
Probability	0.36	0.31	0.1	

   a) Find the probability of choosing a brown sock.
   The experiment is carried out 300 times.
   b) Estimate the number of times a white sock is chosen.